p53 - A Guardian of the Genome and Beyond

*Edited by Mumtaz Anwar,
Zeenat Farooq, Mohammad Tauseef
and Vijay Avin Balaji Ragunathrao*

Published in London, United Kingdom

IntechOpen

Supporting open minds since 2005

p53 - A Guardian of the Genome and Beyond
http://dx.doi.org/10.5772/intechopen.91087
Edited by Mumtaz Anwar, Zeenat Farooq, Mohammad Tauseef and Vijay Avin Balaji Ragunathrao

Contributors
Kiyoto Kamagata, Salvatore Raimondo, Tommaso Gentile, Luigi Montano, Mariacira Gentile, Bakhanashvili Mary, Yusuf Tutar, Kubra Acikalin Coskun, Merve Tutar, Asiye Gok Yurttas, Elif Cansu Abay, Nazlıcan Yurekli, Bercem Yeman Kiyak, Kezban Uçar Çifçi, Mervenur Al, Mumtaz Anwar, Zeenat Farooq, Vijay Avin Balaji Ragunathrao, Rakesh Kochhar, Shahnawaz Wani

Notice
Statements and opinions expressed in the chapters are these of the individual contributors and not necessarily those of the editors or publisher. No responsibility is accepted for the accuracy of information contained in the published chapters. The publisher assumes no responsibility for any damage or injury to persons or property arising out of the use of any materials, instructions, methods or ideas contained in the book.

First published in London, United Kingdom, 2022 by IntechOpen
IntechOpen is the global imprint of INTECHOPEN LIMITED, registered in England and Wales, registration number: 11086078, 5 Princes Gate Court, London, SW7 2QJ, United Kingdom
Printed in Croatia

British Library Cataloguing-in-Publication Data
A catalogue record for this book is available from the British Library

Additional hard and PDF copies can be obtained from orders@intechopen.com

p53 - A Guardian of the Genome and Beyond
Edited by Mumtaz Anwar, Zeenat Farooq, Mohammad Tauseef and Vijay Avin Balaji Ragunathrao
p. cm.
Print ISBN 978-1-83968-145-5
Online ISBN 978-1-83968-146-2
eBook (PDF) ISBN 978-1-83968-147-9

We are IntechOpen,
the world's leading publisher of
Open Access books
Built by scientists, for scientists

6,000+
Open access books available

146,000+
International authors and editors

185M+
Downloads

Our authors are among the

156
Countries delivered to

Top 1%
most cited scientists

12.2%
Contributors from top 500 universities

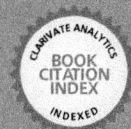

Interested in publishing with us?
Contact book.department@intechopen.com

Numbers displayed above are based on latest data collected.
For more information visit www.intechopen.com

Meet the editors

Mumtaz Anwar, Ph.D., is currently an assistant professor (research) at the Department of Pharmacology and Regenerative Medicine, University of Illinois Chicago, USA. Dr. Anwar obtained his Ph.D. in Cancer Biology and Molecular Epigenetics at the Faculty of Medicine, Department of Gastroenterology, PGIMER, Chandigarh, India. His Ph.D. thesis concerned the investigation of the Wnt signaling pathway in search of new tumor prognostics and biomarkers in colorectal cancer. After completing his Ph.D., he studied epigenetic approaches and other molecular cell signaling mechanisms (in vitro and in vivo) in the context of vascular biology of the lungs at the University of Illinois Chicago. He is a member of various scientific organizations and societies including the American Heart Association (AHA), American Society for Pharmacology and Experimental Therapeutics (ASPET), North American Vascular Biology Organization (NAVBO), and the United States and Canadian Academy of Pathology (USCAP). He is the recipient of various awards including the Outstanding Young Investigator Travel Award.

Zeenat Farooq, Ph.D., is currently a postdoctoral research associate at the Department of Medicine, University of Illinois Chicago. Dr. Farooq obtained her Ph.D. from the School of Biotechnology, University of Kashmir, India. The focus of her Ph.D. was the elucidation of protein factors that play a role in the organization and maintenance of heterochromatin and regulation of transcriptional gene silencing. She has worked actively on various aspects of epigenetics and has authored many papers in reputed journals. She moved to the University of Illinois Chicago in 2020 and studied the role of regulation of mRNA translation. Her recent work is focused on understanding the role of a novel hexokinase in altered glucose metabolism and predisposition to Alzheimer's disease. Dr. Farooq has authored one book and many book chapters. She is a member of various scientific organizations and societies including the American Heart Association (AHA).

Mohammad Tauseef, Ph.D., is currently an associate professor at the Department of Pharmaceutical Sciences, Chicago State University, Illinois, USA. Dr. Tauseef obtained his Ph.D. in Cardiovascular Pharmacology at the Department of Pharmacy, University of Delhi, India. His Ph.D. thesis concerned the investigation of inflammatory cell signaling molecules, with specific emphasis on cardiovascular pharmacology. He is the author of many book chapters and journal articles on molecular endothelial and cardiovascular biology and is involved in other scientific activities. He is a member of various scientific organizations and societies, including the American Heart Association (AHA). He is the recipient of various awards including an AHA postdoctoral fellowship, Midwest Affiliate, Outstanding Young Investigator Travel Award, Research Recognition award, Best Teacher award, and other postdoctoral competition category awards.

Vijay Avin Balaji Ragunathrao, Ph.D., is currently an assistant professor (research) in the Department of Pharmacology and Regenerative Medicine, University of Illinois Chicago, USA. He received his Ph.D. in Biotechnology from the Department of Biotechnology, Kuvempu University, India. His Ph.D. thesis emphasized the chemo-preventive effects of herbal-based drugs and small molecule inhibitors in the inhibition of angiogenesis. Dr. Ragunathrao began his postdoctoral career studying the vascular pathobiology of the lungs at the University of Illinois Chicago. Currently, his research focuses on the Signaling of GPCRs and tyrosine kinase receptors to induce tumor angiogenesis. He has published several articles in high-impact journals. He is also a reviewer and editor of various scientific journals. As an independent researcher, his interests include tumor angiogenesis and lung vascular homeostasis.

Contents

Preface

Genetic information within the cell is contained and stably inherited in the form of deoxyribonucleic acid (DNA). To ensure faithful transmission of this genetic information, it is important for the cell to accurately copy this information. This is followed by the division of old cells into daughter cells. Both processes are essential to maintain the integrity and functionality of the cells. Also, properly controlled cell division is essential to maintain cells in a healthy state, and perturbances in this process lead to the transformation of healthy cells into malignant ones. One of the ways through which the formation of new cells is properly controlled is regulation with the help of specialized proteins, such as p53. This protein plays its cell protective role under the most widely studied conditions and this characteristic lends it its name of "the guardian of the genome." Additionally, p53 is also involved in an array of other functions. With advancements in our knowledge due to the development of new scientific techniques, we have come to appreciate many more roles of this protein, ranging from the prevention of cancer to its role as an environmental biomarker.

This book highlights p53's vast array of functions in a cell, including its lesser-known roles. It is divided into three sections. Section 1 includes an introductory chapter (Chapter 1) on p53. Section 2 includes chapters on the role of p53 in human cancers (Chapter 2), in DNA repair (Chapter 3), and in gene regulation and gene therapy (Chapter 4). Section 3 includes a chapter on the role of p53 as an environmental biomarker (Chapter 5) and a chapter on the study of p53 at a single molecule level (Chapter 6), revealing the dynamics and energetics of p53 binding to DNA.

This book answers some of the most fundamental as well as some of the most obscure questions about p53. We hope it elicits interest in research to uncover and shed light on other uncharacterized functions of this protein.

Mumtaz Anwar
Department of Pharmacology and Regenerative Medicine,
University of Illinois at Chicago,
Chicago, USA

Zeenat Farooq
Postdoctoral Research Associate,
Division of Endocrinology and Metabolism,
Department of Medicine,
University of Illinois at Chicago,
Chicago, USA

Mohammad Tauseef, MPharm Ph.D.
Associate Professor,
Department of Pharmaceutical Sciences,
Chicago State University,
Chicago, USA

Vijay Avin Balaji Ragunathrao, Ph.D.
Assistant Professor (Research),
Department of Pharmacology and Regenerative Medicine,
University of Illinois at Chicago,
Chicago, USA

Section 1

Introduction

Introductory Chapter: p53 - The Miracle Protein That Holds the Distinction of Being "Guardian of the Genome"

Zeenat Farooq and Mumtaz Anwar

1. Introduction

P53 is a protein encoded by TP53 gene in humans. This gene is located on the short arm of chromosome 17 in humans [1]. The gene contains 11 exons and several regulatory regions. The gene is highly conversed in nature and is found across invertebrate and vertebrate species. However, there is a high degree of variability in the coding sequence of p53 in vertebrate and invertebrates. The protein encoded by TP53 is typically known as p53 because in earlier days (around 1979), it appeared to localize at around 53 KDa on a sodium dodecyl sulphate–polyacrylamide gel electrophoresis (SDS-PAGE) gel. However, it was later found that the protein is smaller in size and the lag in migration in the gel occurred due to the abundance of proline residues that cause a *kink* in the structure. The actual mass of the protein, based on summation of molecular masses of all the amino acid contained is around 43.7KDa [2]. Many terms are used for the identification of p53-like tumor protein p53, tumor suppressor p53, phosphoprotein p53, and so on. By far, the most significant working definition offered by any term for its identification is *p53, the guardian of the genome*. This term inherits its *"guardian status"* by the fact that p53 plays a crucial and quintessential role in guarding (protecting) the genome against damage and is therefore found to be mutated in many forms of cancer. In fact, it holds the title of being the most frequently mutated gene in all cancers, documented to be mutated in more than 50% of all cancers [3]. The protein performs its guardian role by acting as a transcription factor and regulating the expression of various genes.

In its three-dimensional structure, p53 protein consists of the following domains, briefly described from N to C terminus below (**Figure 1**) [4–6].

- N-terminus transcription activation domain (TAD) or activation domain 1 (AD1). It is rich in acidic residues. It plays role in regulation of pro-apoptotic genes.

- Activation domain 2 (AD2). It is important for apoptotic activity of p53.

- Proline-rich domain. It is responsible for lag in migration on SDS-PAGE.

- DNA-binding domain (DBD). It plays role in binding to DNA elements on target genes.

- Nuclear localization signal (NLS). It consists of a group of amino acids that are involved in localization of the protein into the nucleus through nuclear pore.

- Self-oligomerization domain (OD). Oligomerization is important for self-annealing and activity of p53.

- C-terminal domain that antagonizes the function of DNA-binding domain.

In its ground state, p53 exists inside the cells in the form of a complex with another protein mdm2 (HDM2 in humans). This dimeric association holds p53 in an inactive state. Mdm2 is also a ubiquitin ligase, which ubiquitylates p53 and marks it for proteolytic degradation. In this manner, p53 undergoes a continuous turn-over in the cells, with a half-life of about 20 minutes. Upon activation, p53 dissociates from mdm2 and becomes available to contribute to a myriad of cellular functions. It exists as a tetramer in its active state. The most common mechanism of p53 activation is phosphorylation at multiple residues.

Upon activation of a stress-signaling cascade in a cell-like DNA damage, activation of proto-oncogenes, or apoptotic pathways, p53 becomes phosphorylated by a variety of kinases, each activated by a particular type of stress signal. Phosphorylation of p53 brings about a conformational change in the protein that interferes with its binding to mdm2 and instead promotes oligomerization of p53. Afterward, p53 moves into the nucleus with the help of NLS and binds to its target genes to promote their transcription. The kinase enzymes therefore favor p53 function in two ways (**Figure 2**).

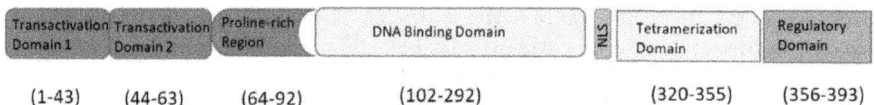

Figure 1.
Domain structure of p53 showing its various domains and their relative size from N to C terminal.

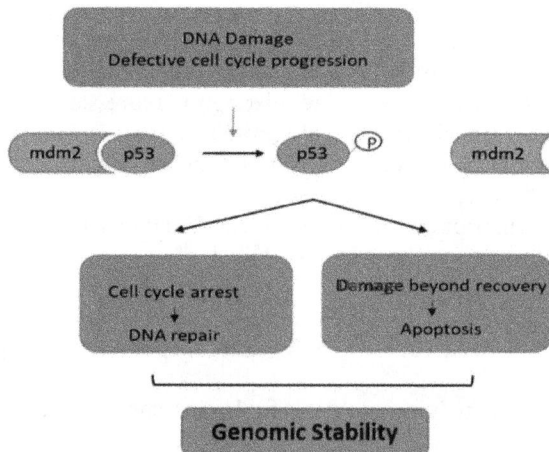

Figure 2.
Outline model depicting the effect of p53 on cells upon activation, leading to cell cycle arrest and DNA repair. If the repair fails, p53 activates pro-apoptotic genes to embark the cells on the path of apoptosis.

- Increase the half-life of p53 by promoting dissociation from mdm2. This increases the cellular concentration of p53 to make it available for the challenge in hand.

- Phosphorylation event favors self-oligomerization, which is essential for p53 activity.

p53 kinases fall into two major groups. Additionally, oncogenes can also activate p53.

- Kinases belonging to MAPK family. These include p38 MAPK, JNK1-3,ERK1-2. These are activated in response to stresses such as membrane damage, oxidative stress, heat shock, osmotic shock, etc.

- Kinases belonging to ATM family. These include ATM, ATR, CHK1, CHK2, DNA-PK. These are activated as a result of cell cycle checkpoint responses induced due to DNA damage.

2. Cellular roles of p53

The quintessential roles of p53 within the cells are as follows [7].

2.1 DNA damage and repair

Upon sensing DNA damage as a result of genotoxic insults, kinases such as ATM and ATR become activated and phosphorylate p53. p53, in turn, activates transcription of proteins that lead to cell cycle arrest at G1/S phase. This allows enough time for the DNA repair proteins to repair the damaged DNA. This process ensures that damaged DNA does not replicate and become inherited by daughter cells through cell division. Once the repair is complete, the cell goes back to the unstimulated state and starts diving normally.

2.2 Apoptosis

The term apoptosis refers to programmed cell death. It is a process by which damaged cells undergo a carefully orchestrated signaling program that culminates in death of the cells without harming neighboring healthy cells of the tissue. This phenomenon occurs when a cell accumulates damage to such an extent that repair is not possible. p53 plays a very critical role in initiating apoptosis of such cells. Both processes are interrelated.

Because of the central role played by p53 in maintaining cellular homeostasis and genome integrity, mutations in the gene are detrimental for p53 function. A large number of mutations have been identified in the gene, which result in the formation of a mutant p53 protein that no longer retains its DNA-binding or oligomerization ability, leading to loss of function. Some mutations have been observed in the DNA-binding domain, which affect binding of p53 to its target genes. Other mutations in the oligomerization prevent p53 sub-units from coming together and forming a functional, oligomeric transcription factor. Another aspect of such mutations is that a single-mutant p53 subunit can prevent oligomerization of wild-type subunits, exerting a dominant negative effect. All these mutations have been identified in many forms of cancer. Additionally, p53 promoter has been shown to undergo an increase in promoter methylation, which leads to decrease in its expression. This mechanism of epigenetic regulation of p53 expression was first of all demonstrated by Bird et al. and has ever since been observed in various other forms of cancer [8–10]. The phenomenon of increase in promoter DNA methylation to decrease expression of cognate gene is also identified as a key epigenetic mechanism with a wide array of cellular functions [11]. According to some reports, it is not only the dissociation of p53 from mdm2 which increases its half-life and cellular availability but some signaling cascades stimulate the translation of p53 mRNAs to increase cellular levels. Increase in mRNA translation of p53 has also been observed to take place in stem cells to trigger differentiation [12]. With the availability of better techniques to carry out research, more exciting work on p53 is being carried out and published, which sheds light on newer and exciting functions of p53.

This book focuses on the roles of p53 as a guardian of genome, explaining in detail various roles performed by the protein under different physiological conditions. The following chapters talk at length about different facets of p53, each related to its cell protective function in light of both established phenomena and latest research in the field on p53.

Author details

Zeenat Farooq and Mumtaz Anwar*
Department of Pharmacology and Regenerative Medicine, College of Medicine, University of Illinois at Chicago, Chicago, IL, USA

*Address all correspondence to: mumtazan@uic.edu; mumtaz_anwar1985@yahoo.co.in

IntechOpen

References

[1] Isobe M, Emanuel BS, Givol D, Oren M, Croce CM. Localization of gene for human p53 tumour antigen to band 17p13. Nature. 1986;**320**(6057):84-85

[2] Ziemer MA, Mason A, Carlson DM. Cell-free translations of proline-rich protein mRNAs. The Journal of Biological Chemistry. 1982;**257**(18):11176-11180

[3] Hollstein M, Rice K, Greenblatt MS, Soussi T, Fuchs R, Sørlie T, et al. Database of p53 gene somatic mutations in human tumors and cell lines. Nucleic Acids Research. 1994;**22**(17):3551-3555

[4] Venot C, Maratrat M, Dureuil C, Conseiller E, Bracco L, Debussche L. The requirement for the p53 proline-rich functional domain for mediation of apoptosis is correlated with specific PIG3 gene transactivation and with transcriptional repression. The EMBO Journal. 1998;**17**(16):4668-4679

[5] Larsen S, Yokochi T, Isogai E, Nakamura Y, Ozaki T, Nakagawara A. LMO3 interacts with p53 and inhibits its transcriptional activity. Biochemical and Biophysical Research Communications. 2010;**392**(3):252-257

[6] Harms KL, Chen X. The C terminus of p53 family proteins is a cell fate determinant. Molecular and Cellular Biology. 2005;**25**(5):2014-2030. DOI: 10.1128/MCB.25.5.2014-2030. 2005

[7] Levine AJ. p53, the cellular gatekeeper for growth and division. Cell. 1997;**88**(3):323-331

[8] Malhotra P, Anwar M, Nanda N, Kochhar R, Wig JD, Vaiphei K, et al. Alterations in K-ras, APC and p53-multiple genetic pathway in colorectal cancer among Indians. Tumour Biology. 2013;**34**(3):1901-1911

[9] Bird AP. CpG-rich islands and the function of DNA methylation. Nature. 1986;**321**(6067):209-213

[10] Farooq Z, Shah A, Tauseef M, Rather RA, Anwar M. Evolution of Epigenome as the Blueprint for Carcinogenesis. Rijeka: IntechOpen; 2021. DOI: 10.5772/intechopen.97379

[11] Maimets T, Neganova I, Armstrong L, Lako M. Activation of p53 by nutlin leads to rapid differentiation of human embryonic stem cells. Oncogene. 2008;**27**(40):5277-5287

[12] Hollstein M, Sidransky D, Vogelstein B, Harris CC. p53 mutations in human cancers. Science. 1991;**253**(5015):49-53

Section 2

Role of p53 in Human Cancers, DNA Repair, Gene Regulation and Gene Therapy

Chapter 2

Role of p53 in Human Cancers

Kubra Acikalin Coskun, Merve Tutar, Mervenur Al,
Asiye Gok Yurttas, Elif Cansu Abay, Nazlican Yurekli,
Bercem Yeman Kiyak, Kezban Ucar Cifci and Yusuf Tutar

Abstract

TP53 codes tumor protein 53-p53 that controls the cell cycle through binding DNA directly and induces reversible cell-cycle arrest. The protein activates DNA repair genes if mutated DNA will be repaired or activates apoptotosis if the damaged DNA cannot be fixed. Therefore, p53, so-called the "guardian of the genome," promote cell survival by allowing for DNA repair. However, the tumor-suppressor function of p53 is either lost or gained through mutations in half of the human cancers. In this work, functional perturbation of the p53 mechanism is elaborated at the breast, bladder, liver, brain, lung cancers, and osteosarcoma. Mutation of wild-type p53 not only diminishes tumor suppressor activity but transforms it into an oncogenic structure. Further, malfunction of the *TP53* leads accumulation of additional oncogenic mutations in the cell genome. Thus, disruption of *TP53* dependent survival pathways promotes cancer progression. This oncogenic *TP53* promotes cell survival, prevents cell death through apoptosis, and contributes to the proliferation and metastasis of tumor cells. The purpose of this chapter is to discuss the contribution of mutant p53 to distinct cancer types.

Keywords: p53, TP53, mutation, loss-of-function, breast cancer, bladder cancer, liver cancer, brain cancer, osteosarcoma

1. Introduction

Cancer is a disease that occurs as a result of mutations in the genes responsible for the DNA repair, cellular proliferation, and cell cycle checkpoints, resulting from the unbalanced equilibrium of oncogenes and tumor suppressor genes that cause uncontrolled growth and invasive migration of the cells [1]. In healthy cells, DNA damage can be repaired by distinct DNA repair mechanisms and the cell can continue to its normal functions. However, if the repair mechanism is perturbed, cells can not correct the changes caused by mutations. In spite of this, the protein product of this gene can be degraded or during proliferation, checkpoints in the cell cycle detect the mutation and the cell undergoes apoptosis [2]. However, cancer cells are master to inactivate the cell cycle checkpoints by mutations on tumor suppressor genes and to activate tightly regulated proto-oncogenes. Proto-oncogenes are expressed only when required [3]. They are expressed in a controlled manner for cell growth and act as mitogens in healthy cells [4]. Due to their mitogenic roles, most of the mitogenic genes within the genome are upregulated in the case of cancer, and most of these genes are considered as proto-oncogenes. As a result of accumulated mutations on proto-oncogenes, the cell enters an uncontrolled division pathway [3]. Further at some point, accumulation of mutations in DNA repair mechanisms and tumor suppressor genes suppress cell death mechanisms in tumor cells and oncogenes are upregulated and over-activated in tumor cells. All of these changes cause loss of cell cycle checkpoints to control and DNA repair mechanism's function, and the cell eventually is transformed into a cancer cell [5].

The *TP53*, which is known as the guardian of the genome and is one of the proteins that play the most important role in the cell cycle, was first noticed in animal experiments in 1979 when the tumor tissues were examined [6]. p53, a short-lived protein synthesized by the *TP53* gene in cells, was named "p53," taking its name from its molecular weight of 53 kDa (kilodalton) [7]. p53 is a transcription factor that regulates cell division. Specifically, p53 functions at cell differentiation and initiation of DNA repair mechanism, and is a protein that has a role in suppressing cancer in several organisms [8]. The principal mechanism can be summarized with the understanding that p53 is not always active in typical cells and their activity is minimal in the case of healthy cells. p53 protein is activated only after DNA damage.

There are two important steps in the p53 activation process. In the first step, the half-life of p53 increases dramatically, which means the amount of functional p53 increases and degradation of p53 decreases in the cell, then it is observed that p53 proteins rapidly accumulate within the cell due to the DNA damage as illustrated in **Figure 1**. Thereafter, conformational changes convert the protein into transcriptional regulatory protein form through phosphorylation and enable p53 to be functionally activated. Thus, the increased amount of functional p53 activates DNA repair mechanisms. Normally, when the cells have no DNA damage, the amount of p53 is kept at a low level by protein degradation.

A protein called MDM2 (the murine double minute 2) interferes with p53 and inhibits the function of p53 and sends p53, which function in the nucleus, from the nucleus to cytosol. MDM2 also works as a ubiquitin ligase (**Figure 2**). This function of MDM2 helps the destruction of functional p53 by sending p53 to the ubiquitin proteasomal system (UPS), and the amount of p53 in the cell is reduced [1, 9, 10]. When genomic damage occurs in cells, cell growth halts, p53 stimulates

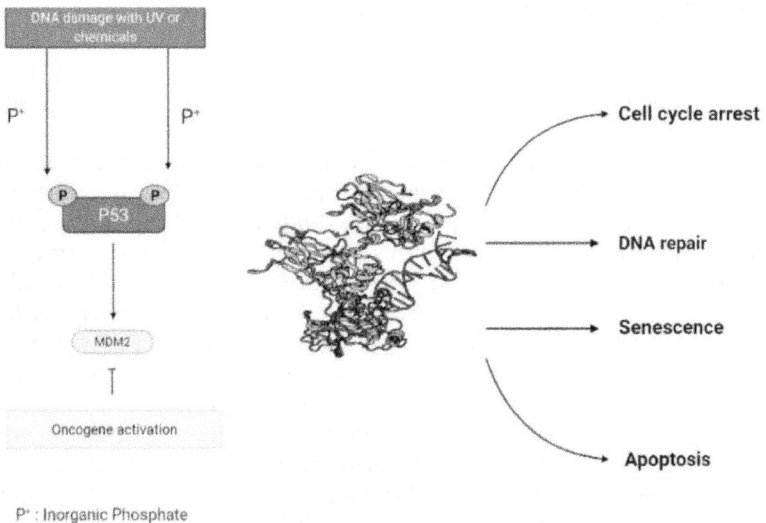

P⁺ : Inorganic Phosphate

Figure 1.
p53 activation summary; DNA damage enables p53 to become active by inhibition of MDM2 that results in cell cycle arrest, apoptosis, senescence, and DNA repair.

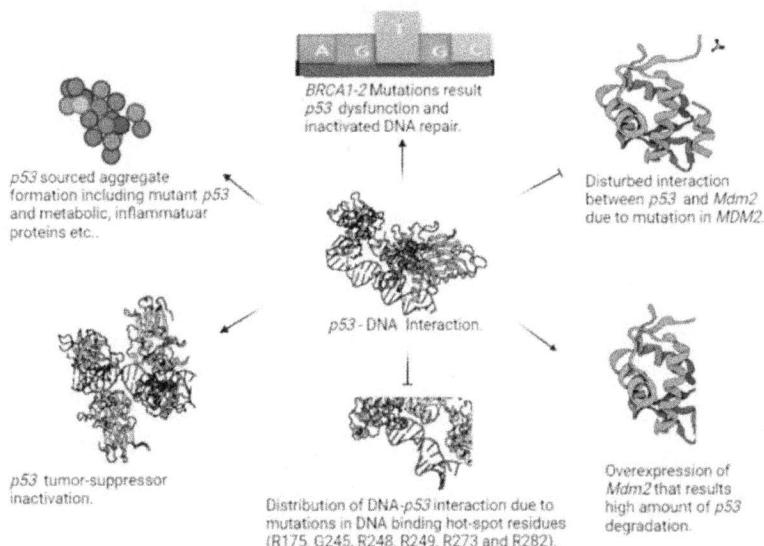

Figure 2.
Overview of inactivation of p53 with distinct mechanisms in breast cancer (MDM2 PDB ID: 1T4F; p53 PDB ID: 1TUP and p63 PDB ID: 3US1).

programmed cell death-apoptosis [11]. Due to its cancer suppression ability, cancer cells adapted to inhibit p53 function in different ways and escape from senescence and apoptosis by distinct mechanisms [10].

These features and changes on the *p53* gene contribute to cancer transformation via escaping from the cell cycle checkpoints and cell death. Therefore, *p53* mutations are crucial for most of the cancer cells to sustain their existence. Observation of high-frequency *p53* mutations in most of the cancerous cell types can be explained in this way [7]. However, it is known that each type of cancer follows different adaptations and genomic rearrangements depending on specific alterations and environmental factors. P53 mutations and functions also change according to the cancer types with distinct mechanisms. This review elaborates on these distinct mechanisms of *p53* mutations in different cancer types.

1.1 Breast cancer

Breast cancer is considered as one of the most frequent types of cancer [12, 13]. Breast cancer morbidity and mortality rates are higher nowadays. There are many different treatment approaches for breast cancer [14]. However, breast cancer in different patients has a variety of symptoms, disease progression, and drug response which proved that breast cancer subtypes are distinct and need different treatment regimens. Breast cancer has a heterogenic nature. Thus, heterogeneity creates different clinical features in the cancer cells [15]. Breast cancer can show differences in the expression of the hormonal receptor as the result of different genetic alterations and rearrangements within the cell [16]. These differences cause different subtypes of breast cancer that show different strategies to survive. With the help of gene expression analysis (genome sequencing, transcriptional and translational analysis, etc.), luminal ER-positive (luminal A and luminal B), HER2 enriched, and triple-negative (basal-like) types are identified as three major types of breast cancer [17].

Mutant *p53* plays a pivotal role in the prognosis of approximately 23% of breast cancer [18]. TP53 mutations are the most common genetic modifications in breast carcinomas, according to recent next-generation sequencing-based research, accounting for 30% of them. On the other hand, the distribution of these mutations is strongly associated with tumor subtypes. In 26% of luminal tumors (17% of luminal A, 41% of luminal B), 69% of molecular apocrine tumors, and 88% of basal-like carcinomas, mutations have been elucidated [19]. Further, protein kinases such as CHK1, CHK2 (Rad53), ATM (ataxia-telangiectasia mutated), and ATR (Rad53-related protein), which respond to DNA damage sentinels, such as BRCA1, also control p53 activity and stability. The kinases directly phosphorylate p53, affecting its instability and function [20].

Although the general prevalence of p53 mutation in breast cancer is around 20%, specific forms of the cancer are associated with greater rates (**Figure 2**). A number of studies, for example, have found an elevated rate of p53 alterations in malignancies caused by carriers of germline BRCA1 and BRCA2 mutations. Surprisingly, p53 mutation occurs in 100% of instances of typical medullary breast carcinomas. This is particularly interesting because it is now well accepted that medullary breast tumors exhibit clinicopathological characteristics with BRCA1-associated instances. Furthermore, methylation-dependent BRCA1 silencing is frequent in medullary breast tumors [21].

TP53 mutation is found in nearly half of HER2 amplified malignancies [13]. The type of change is clearly linked to the breast cancer subtype, with a higher frequency of substitutions in luminal tumors, resulting in a p53 protein with possible novel functionalities such as p63 inactivation. p63 is a member of the p53 family that also has a tumor suppressor activity [22]. The majority of mutations focused on missense mutations. The most frequent missense mutations in p53 are located within the DNA binding domain. Especially in six frequent "hotspot" amino acid codons (R175, G245, R248, R249, R273, and R282) (**Figure 2**) [23].

Some mutant *p53* in the cancer cells lose its tumor-suppressive activity of the wild-type *p53* and shows strong oncogenic functions, defined as a gain of function that provides a selective advantage during tumorigenesis progression [24]. Most of the p53 mutations are seen in the DNA binding domain that allows the expression of DNA repair system proteins [25].

Also, due to mutations, p53 can act like prions and cause accumulation within the cancer cells by binding other proteins, such as metabolism, RNA processing, and inflammatory response [18]. On the other hand, deregulation of MDM2-p53 pathway due to amplification and overexpression of *MDM2* oncogene which is a master regulator of the *p53* tumor suppressor activity, and mutations or deletions of *p53* has been correlated to the initiation, progression, and metastasis of breast cancer [10].

Mutations in TP53, as well as the deletion of RB1 and CDKN2A, are among the most well-known genetic changes in tumor suppressor genes in basal-like breast cancer and triple-negative breast cancer [20]. Indeed, up to 80% of basal-like breast cancer have TP53 alterations, which include nonsense and frameshift mutations. The RNA of 99.4% of basal-like breast cancer patients had TP53 mutant-like status. TP53 mutations may have varied effects depending on the breast tumor subtypes. There is now evidence that inactivation of *p53* by mutation, amplification of MDM2 or MDM4, or infrequent alterations in other p53 pathway components causes luminal cancers [26].

In conclusion, the activity of p53 can be inhibited by either mutation in *p53* or mutations in *p53*-interacted proteins that regulate its function.

1.2 Bladder cancer

Up to 50% of cancer cases have acquired a mechanism that inactivates p53 function to bypass apoptosis. The most indisputable fact about p53 is its high frequency of modifications in human cancer. Mutant p53 proteins form a complex family of several 100 proteins with heterogeneous properties. The *p53* tumor suppressor gene located on chromosome. 17p13 is one of the most frequently mutated genes in all human malignant diseases, including bladder cancer [27].

It is known that p53 gene mutations occur early in the pathogenesis of bladder cancer and late in other cancer types [28, 29]. Tumor protein *p53* gene mutation is an important marker for bladder cancer progression and is associated with poor prognosis and recurrence [30]. The *TP53* gene is responsible for maintaining genome integrity as it encodes a protein that is activated in response to cellular stress to repair possible DNA damage [31].

About 60% of bladder cancer cases result in *mutp53* (mutant-p53) in exon 5–11. *mutp53* is commonly associated with the *mutRb* gene in high-grade, invasive, and poorly prognostic bladder cancer [32]. Up to 20% of all BC cases were caused by the p53 gene mutation in exons 1–4, accompanied by *mutCDKN2a* and loss of ARF function. Therefore, it has been suggested that mutations in the *RB, CDKN2a,* and *ARF* genes may follow the *p53* mutation [33].

1.2.1 p53 mutations in bladder cancer

The *p53* gene is mutated in 20–60% of bladder tumors. Especially codon 80 and codon 285 are the regions where mutations are the most common. The gene encodes *p53* has a conserved sequence and has 5 polymorphisms that are located in coding part of the gene. While four of them are codon 34, 36, 47, 72 in exon 4; one was found in exon 6 codon 213. Most of the polymorphisms in *p53* were found in the intronic region. There are two in intron 1, one in intron 2, one in intron 3, two in intron 6, five in intron 7, and one in intron 9. Of these, polymorphisms at codon 72 and codon 47 are well characterized [34].

Codon 280 and 285 in exon 82 are hot regions for mutation formation. Codon 280 is common in 1.2% of all cancer types and mutant in 5.1% of urinary bladder cancers. These values are 0.82% of all cancer types for codon 285, compared to 4.3% of urinary bladder cancers [35].

1.2.2 p53 polymorphisms in bladder cancer

The incidence of the codon 72 arginine/proline (Arg-CGC/Pro-CCC) polymorphism varies by ethnic group and geography [36]. The region containing the five repeat pxxp sequence (proline) located between amino acids 61 and 94 in *p53* is thought to be involved in the signal transduction of this motif through its binding activity to the SH3 region. In cell culture studies, defects in the suppression of tumor cell growth by *p53* have been associated with the deletion of the proline-rich region. Conversion of the G base to the C base causes the conversion of arginine AA at codon 72 to proline AA. The Arg carrying a form of *p53* was found to be significantly more associated with tumor growth than the proline carrying form [37]. In a study, it was shown that the Arg/Arg genotype increases the risk of developing bladder cancer [38]. In addition, Kuroda et al. found an increased risk of urethral cancer in smokers with the Pro/Pro genotype [39].

Silent mutations at codon 36 (CCG → CCT); It was observed that *MDM2* decreased the affinity of TP53 mRNA and decreased the activity of *P53* in apoptosis.

Three similar polymorphisms, D21D (GAC → GAT), P34P (CCC → CCA), and P36P (CCG → CCA), are found in key regions in MDM2-binding TP53 mRNA. According to the latest findings, translation inhibition is inhibited by microRNA (miRNA) targeting gene coding sequences [40, 41].

1.3 Brain cancer

There are more than a 100 different types of brain tumors which are either primary brain tumors that arise from the central nervous system (CNS) cells or secondary brain tumors that have metastasized from other tissues in the body. While primary brain tumors make up about 2% of all cancers, secondary brain tumors are seen 10 times more often.

Brain tumors can be considered as a heterogeneous group of benign and malignant tumors. Even though most types are cancerous, benign tumors can also become damaging for the brain tissue. Their classification using various parameters and a grading system (I–IV) by the World Health Organization (WHO) is a helpful criterion when choosing the best approach in diagnosis and treatment. When classifying brain tumors, in addition to histological criteria, molecular genetic alterations are also taken into consideration and nomenclatured accordingly [42, 43].

Meningiomas, originating in the dura, are usually benign and can be removed by surgery; they represent around 36% of all primary brain tumors [43]. Almost 75% of malignant primary tumors and 29% of all brain tumors are gliomas. They originate from glial cells and are grouped as circumscribed (grade I) and diffusely infiltrating (grades II, III, and IV) gliomas. Circumscribed gliomas, called ependymomas, are usually benign and can be cured with complete resection. They make up about 7% of gliomas and mostly affect children. The latter group, including astrocytomas (about 75% of gliomas) and oligodendrogliomas (about 6% of gliomas), are usually malignant and difficult to cure. This group also includes mixed gliomas which are not easy to diagnose as the composition of cell type, whether astrocytes or oligodendrocytes, may not be accurately determined [42–44]. As the most common and deadly primary tumor, glioblastoma makes up almost half of all gliomas and about 80% of malignant gliomas. About 30% of glioblastomas have *p53* mutations related to loss or gain-of-function, and also dominant-negative effects [45].

One of the most studied proteins, *p53* is best known for its tumor suppressor role. In cases of tumor stress, it stops the cell cycle to either let DNA repair itself or cause cell death with interferes with tumorigenesis. Its involvement plays a major role in the regulation of apoptosis and therefore cases of *p53* mutations lead to deregulation and dysfunction of apoptotic responses through *p53*-dependent mechanisms. It is already one of the most common mutant genes in human cancers, but it is also known to be closely involved with cancers related to CNS, and also other neurological diseases including Alzheimer's disease, Parkinson's disease, and Huntington's disease [46, 47]. Studies done with transgenic mice overexpressing amyloid-β have demonstrated increased expression and accumulation of *p53* in the brain, which was also seen in the brains of Alzheimer's patients [48, 49].

p53 has also been of great importance during the development of the brain and regulation of neuroinflammation [47, 50]. One of the earliest studies performed on *p53*-deficient mice has demonstrated abnormal brain development. As a result of decreased apoptosis, defects in the closing of the neural tube have occurred. This disruption has eventually led to exencephaly followed by anencephaly [51].

Inactivation of *p53* happens through several mechanisms including the disruption of its gene expression or protein stability and also loss or mutation of the gene itself. These mechanisms result in malignant properties such as invasiveness, undifferentiated status, and genetic stability. The frequency of *p53* mutations depends on the type of tumor. Glioblastoma, the most lethal one, has the highest incidence of 70%. Mixed gliomas and astrocytomas are moderate, 40%, and 50%, respectively. Oligodendrogliomas have the lowest incidence among all gliomas. In general, tumor grades are determinant in the occurrence rate of *p53* mutations, of which missense mutation is the main one. C:G → A:T mutation is the most common mutation of *p53* seen at CpG sites, affecting the DNA binding properties through three codons, R248, R273, and R175, in the DNA binding domain according to The Cancer Genome Atlas (TCGA). Mutations of this domain have led to gain-of-function to induce tumorigenesis. Additionally, splice site mutations, promoter methylations have also been identified [44, 47, 50].

An example of gain-of-function mutation is given in a recent study done on an invasive brain tumor, glioblastoma. Mutation in *TP53* increases the tendency of aggregate formation via mutant *P53* oligomerization due to exposed hydrophobic parts. Once aggregation of this protein takes place in the cell, conditions for cancer initiation and oncogenic activities are likely to be established [52].

Another group analyzed the key genes and pathways of *p53* mutations in low-grade glioma patients. RNA-seq data from the TCGA database were analyzed by various bioinformatics tools to have a deeper understanding of the role of this protein in disease progression. Out of 508 patients, 49% had mutations such as amplification, deletion, truncation, in-frame mutations, and missense mutations throughout the whole gene. Cancer cells with these mutations were then found to be resistant to some chemotherapeutic drugs that are normally used to treat glioma. This is an indication that it is especially important to distinguish whether the patient has *p53* mutation or not to avoid failure of the therapy. In addition, 1100 differentially expressed genes were identified, of which most were associated with pathways related to cancer development and progress [53].

In conclusion, primary brain tumors are difficult to deal with, in terms of understanding their basis and managing the progress. In cases of relevant *p53* mutations, attention can be focused on avoiding the degradation of this protein or using chaperones to reestablish its structural integrity and biological activity. Upstream and downstream molecules can be alternatively targeted to develop other novel therapeutic strategies. Last but not the least, determination of *p53* mutations is a significant step that helps to choose the best individualized therapy for cancer patients.

1.4 Liver cancer (hepatocellular carcinoma)

Liver cancer is the second most common cause of cancer-based mortality worldwide, accounting for 7% of all cancers with 854,000 new diagnoses each year. The main histological subtype of liver cancer is hepatocellular carcinoma (HCC) which originates from hepatocytes. Considering the population, the incidence of hepatocellular carcinoma increases with age, and male individuals are at greater risk. Based on etiological data in HCC; hepatitis virus and HIV infections, smoking and alcohol use, aflatoxin B1 exposure, and metabolic diseases are the factors associated with carcinogenesis. More effective therapies are still being investigated for HCC due to the fact that the methods used in the treatment are less effective, the treatment is accompanied by cirrhosis, liver failure, and the difficulty of grading-staging of the tumor [54, 55].

The functioning of hepatocellular carcinoma induced by carcinogens is caused by multiple dysfunctions on the MDM2-p53 axis. Oncogene activation, genotoxic and ribosomal stress, and hypoxia signals activate the *p53* mechanism. *p53*, the most important tumor suppressor, is also associated with hepatocyte proliferation and metabolism. Hepatitis B virus-X protein (HBx), which binds *p53* and sends it from the nucleus to the cytoplasm, has been shown to play an important role in the development of HCC. Special regions in MDM2 and *p53* are linked to exposure to environmental carcinogens and the development of HCC. Mutations in the MDM2-p53 axis and chronic HCV infection have been shown to trigger the development of HCC [56]. Normally, if MDM2-p53 key regions are not phosphorylated, the increase in MDM2 levels leads to inhibition of *p53* expression activity, which disrupts cell cycle control and stimulates tumor formation. The scientific findings accumulated due to these mechanisms indicate that *p53* is critical for stopping the development of HCC [8, 57].

Clinical case studies suggest that control of *p53* expression for regeneration of liver tissue after partial hepatectomy may regulate CDK2-CDK4 activity, which promotes DNA synthesis in hepatocytes. In addition, in mice with *p53* defects, repair of liver failure and hepatocyte damage is delayed. According to these results; homeostasis of wild-type *p53* expression controls the proliferation and apoptosis of normal hepatocytes. However, mutant *p53* is predominantly a negative inhibitor compared to wild-type *p53*. The fact that mutant *p53* oncogenic potential is a major factor in liver cancer, as with many malignant cancers [8, 54].

The basic mechanism of apoptosis formed by *p53* depends on death signals that directly or indirectly target mitochondria through pro-apoptotic members of the TP53 and Bcl-2 family, both of which have mutations. Healthy liver cells are resistant to *p53*-mediated cell death, and the relationship between mitochondrial translocation of *p53* and apoptosis after DNA damage is rare. In HCC cells, the activation of *p53* encourages stopping the cell cycle instead of apoptosis, and mostly in hepatocytes, the mitochondrial-dependent *p53* apoptosis pathway is blocked. The likely cause of this critical change is the increased expression of hepatic insulin-like growth factor binding protein-1 (IGFBP1), which antagonizes the mitochondrial *p53* pathway and prevents apoptosis as a result of *p53* activation [58, 59].

The main mutation of TP53 in hepatocellular carcinoma occurs in the DNA binding region of *p53*, which causes a lower affinity to bind specific response units of their targeted genes to the array, and *p53*-mediated MDM2 induction decreases. As a result, misregulation of MDM2 results in high levels of mutant *p53* expression in many cancerous cells [58, 60].

The key role of *P53* in tumor development has made *p53* an inspiring target for drug studies that inhibit HCC development. Treatments to restore *p53* function in HCC have been shown to damage cancer cells that express both mutant *p53* and wild-type *p53*. Current treatment approaches for HCC; chemotherapy, radiotherapy, degradation pathways of ADP-ribosylation factor proteins inhibiting *p53*, inhibition of MDM2-p53 connectivity, and the addition of molecules regulating the active region of the *p53* protein [61].

1.5 Osteosarcoma

Osteosarcoma, which can also be called osteogenic sarcoma, is a cancer type that is related to bones. It is a common pediatric bone tumor as it has an annual diagnose rate of 400 children [62]. This type of cancer starts to form when there is a problem with the cells that are responsible to make new bones [62]. Healthy bone cells may have alterations in their DNA, which can result to make new bones when there is no need for them. As a result of making new bones without a need, there will be a cell mass formed with poorly formed bone cells. Then, this cell mass will destroy the body tissue that was healthy in the first place by invading it. Also, as the cancer progress, some cells can spread through the body and metastasize.

There are two kinds of *p53* with different effects on osteosarcoma. Wild-type *p53* functions as a tumor suppressor and the mutant *p53* have a carcinogenic effect and are found to be overexpressed in malignant osteosarcoma [63]. A study proves this overexpression point by using immunochemistry and concluding that mutant *p53* had a 47.7% positive expression rate [63]. On the other hand, since wild-type *p53* is a protein known to be a tumor suppressor, it is expected to have changes due to mutations, etc. in most cancers. With this change process, a response to DNA damage cannot be made and the genome destabilizes. Like other types of cancer, osteosarcoma is also known to have this type of relationship with the *p53* protein. Changes in *p53* are shown to have a correlation with the instability of the genome with osteosarcoma patients [64]. HDM2 is a protein that functions as a negative regulator of *p53* [64]. It is found that if there is an amplification of the HDM2 protein, the expected instability of the genome does not happen. When HDM2 protein amplification happens without mutations happening in *p53* protein, there is not a high level of instability in the genome. When these direct and indirect ways to change *p53* are compared, the alterations that happen with HDM2 amplification do not even correspond to half of the alterations that destabilize the genome caused by a direct mutation in *p53* [64]. So, this implies different ways that cause a change in the *p53* protein happens to create different results. Since this is not a fully established subject, future studies on the different kinds of changes can be found helpful in the research of this disease and its treatments.

TP53 is a gene that works to help assemble the *p53* (or TP53) protein. The prognostic values of osteosarcoma patients with TP53 mutations are also studied. An analysis was made using eight eligible studies which in total had 210 osteosarcoma patients [4]. Final data from this analysis concluded that in two-year survival of osteosarcoma patients, the mutations of TP53 had a negative impact when compared to wild-type ones. So, it is concluded that TP53 mutations are important for the patients' survival rates and are prognostic markers [65]. Although the results from this study conclude that the mutations have an unfavorable impact on survival, there is still a need for larger-scale studies showing three-to-five-year survival of osteosarcoma patients.

The influence of *TP53* mutations is also shown in another study, St. Jude Children's Research Hospital-Washington University Pediatric Cancer Genome Project (PCGP), which concludes that 90% of the patients with osteosarcoma showed a mutation in the *TP53* gene [66]. This study also revealed the type of mutations upon whole-genome sequencing 34 osteosarcoma tumors [66]. They concluded that 55% of the *TP53* mutations are caused by structural variants, and it is found to be second cancer with these types of mutations that is related to the rearrangement of chromosomes instead of point mutations [66]. This effect of TP53 mutations is believed to be the reason for the ineffectiveness of standard doses in radiation therapy.

1.6 Lung cancer

The TP53 gene mutation is one of the most common causes of lung cancer and has a key role in the carcinogenesis of lung epithelial cells. Small cell lung cancer (SCLC) and non-small cell lung cancer (NSCLC) are two main types of lung cancer in humans. Approximately 80% of all lung cancers are NSCLC that creates most of the *TP53* mutations [67].

The TP53 gene has been found in lung cancer pathogenesis with the frequent detection of loss of heterozygosity (LOH) at the location of the *TP53* gene on chromosome 17p13 in lung cancer cell lines and tumor samples. Additionally, it has been shown that the mutations in the TP53 in lung cancers have been linked to a poorer prognosis and increased cellular resistance to therapy [68]. SCLC specimens have the highest prevalence of *TP53* mutations [69]. However, in NSCLC tumor samples, squamous cell carcinomas have the highest frequency of *TP53* mutations and adenocarcinomas have the lowest frequency. The location of *TP53* mutations is mostly in the DNA-binding domain of TP53 and is detected in cancers with and without allele loss at 17p13 [70]. Acquired *TP53* mutations are kept during tumor progression and metastatic spread since *TP53* coding mutations appear early in the evolution of lung cancer and are possibly essential for maintaining the malignant phenotype. Chang and his colleagues clarified that *TP53* mutations were found in 23.2% of primary tumors and 21.4% of metastatic lymph nodes. Moreover, there was 92.9% concordance between 56 patients with NSCLC who had surgical resection in primary tumors and metastatic lymph nodes [71]. This explains that the majority of *TP53* mutations arise before the tumor spreads. They are subsequently preserved throughout the rest of the tumor's development, therefore there is no selection for *TP53* mutations during metastasis [67].

1.6.1 Tobacco-associated lung cancer and TP53 mutations

Tobacco smoking is the major cause of lung cancer, and the risk of lung cancer rises with the number of cigarettes smoked and the length of time spent smoking although 15% of men and 53% of women with lung cancer in the world are never smokers. Furthermore, in the United States and the European Union, tobacco smoking is responsible for more than 90% of lung cancer in males and 74–80% of lung cancer in women [72]. *TP53* mutations are detected in more than half of lung cancers. Therefore, this makes the *TP53* gene one of the most common targets of tobacco smoking-related DNA alterations.

Several studies have previously discovered hotspots on the *TP53* gene, with G:C to T:A (G to T) transversions being a common finding in tobacco-related lung cancer [73]. In addition, 90% of the guanines that undergo these transversion events are found on the non-transcribed DNA strand. There was a lower incidence of G to T transversions in lung cancer tissues from never-smokers than from smokers [74]. Polycyclic aromatic hydrocarbons (PAH) that are found in tobacco smoke are thought to cause the spectrum of G to T transversions. The major metabolite of benzo[α]pyrene which is the most studied member of the PAH class is benzo[α] pyrene diol epoxide (BPDE). Moreover, it is one of the most dangerous carcinogens found in high concentrations of tobacco smoke [75]. A number of studies have demonstrated that BPDE-DNA adduct patterns in the *TP53* gene in bronchial epithelial cells correspond to G to T mutational hotspots at codons 157, 248, and 273. At these codons, G to T transversions are common for bulky adduct-producing mutagens, such as PAHs and BPDE adducts [76].

1.6.2 TP53 mutations in never-smokers and smokers

Several studies have clarified that lung cancer from smokers shows a different and unique mutation spectrum in the *TP53 gene* than lung cancer from never-smokers. Up to 83% of *TP53* mutations were transitioned in female never-smokers with adenocarcinoma patients. On the other hand, *TP53* mutations in smokers were mostly transversions (60%) and deletions (20%). The incidence of TP53 mutations was shown to be proportional to the amount of tobacco smoking in patients with adenocarcinoma [77]. However, never-smokers with adenocarcinoma patients have more mutations in the epidermal growth factor receptor (EGFR) tyrosine kinase than tobacco-associated lung cancer patients and have a higher response to its inhibitors. Additionally, in adenocarcinoma, *TP53* mutations have been found to be closely linked to smokers, while EGFR mutations are statistically substantially more common in females and never-smokers. Moreover, the incidence of K-ras and *TP53* mutations varies between never-smoker lung cancer patients and smoker lung cancer patients [78].

1.6.3 Therapeutic strategies for NSCLC patients with TP53 mutation

TP53 mutations show chemoresistance to lung cancer cells *in vivo* and *in vitro*, according to several studies. If TP53 status is determined, chemo or radiation therapy can be decided. For example, cancers carrying the mutant *TP53* are known to be more resistant to ionizing radiation than tumors containing the wild-type *TP53* [79]. To target the TP53 pathway in cancer, virus-based therapeutic strategies are one of the most advanced strategies. Because *TP53* mutations are common in lung cancer, the treatment with various chemotherapy classes and *TP53* gene replacement techniques has been investigated in both preclinical and clinical settings. When *TP53* gene therapy was studied in lung cancer patients in clinical trials, some researchers have suggested that combining adenovirus (Adp53) gene therapy with chemotherapy medicines and radiotherapy can be effective [80]. For instance, 28 patients with NSCLC were given the *Adp53* gene into their tumors without any other therapy in the phase I clinical trial. Two patients (8%) had a significant reduction in tumor size, and 16 patients (64%) had disease stabilization; the remaining seven patients (28%) had disease progression [81].

There are also several approaches such as rational design and screening of chemical libraries to identify small compounds that target mutant *TP53*. RITA was discovered in the National Cancer Institute's (NCI) drugs that could reduce cell proliferation in a wild-type TP53-dependent way. It reactivates TP53 and promotes apoptosis by breaking the interaction with HDM-2 after attaching to it [82]. As a result, it has been proposed as a crucial drug to target tumors with wild-type TP53 that may be resistant to drugs that restore mutant TP53 activity, such as PRIMA-1 (p53 reactivation and production of large apoptosis). PRIMA-1 that is a low-molecular-weight drug has been discovered to suppress the growth of tumor cells expressing mutant *TP53*. It binds to the core of mutant *TP53*, restoring its wild-type conformation and inducing apoptosis in human tumor cells [83]. A study revealed that although PRIMA-1 did not cause apoptosis in human NSCLC cell lines encoding distinct TP53 proteins, such as A549 (p53wt), LX1 (p53R273H), and SKMes1, it did dramatically impair cell viability (p53R280K). In addition, PRIMA-1 enhances adriamycin-induced apoptosis in A549 and LX1 cells when used in combination with the drug. In a preclinical setting, *Adp53* gene therapy and PRIMA-1 which can restore the transcriptional function of mutant TP53, or RITA, which inhibits MDM2-directed TP53 degradation, have been performed, and some of these techniques are now in clinical development [84]. Last but not least, the combination of the traditional and molecular-targeting cancer treatments with new TP53-based therapeutic methods for NSCLC can offer great potential for targeting only cancer cells.

2. Conclusions

P53 stands at the heart of the cancer mechanism due to its role in cell survival and death. *TP53* essential role in cell fate decision attracts the interest of cancer researchers and makes the protein a superior target for anti-cancer drugs. Therefore, the focus on TP53 research at distinct cancer types increases dramatically and TP53 is targeted by drug designers to inhibit its mutant protein function. *P53*

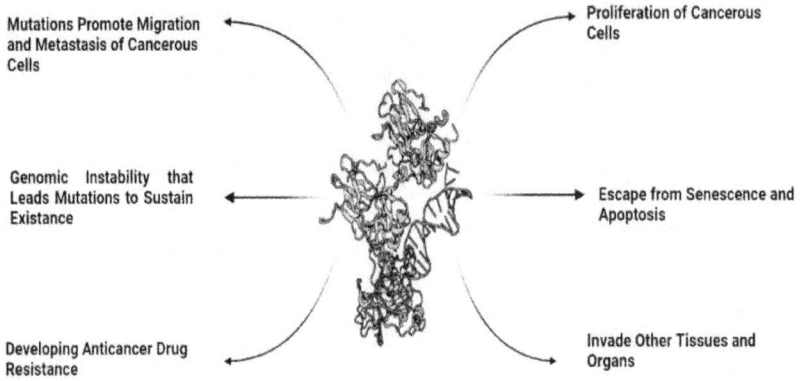

Figure 3.
Roles of p53 in cancerous cells.

and its partner proteins like its negative regulator MDM2 are of further interest for this purpose. This protein-protein interaction features specific properties for allosteric protein inhibition. Yet, the mutant composition of *p53* alters among distinct cancer types. As it is illustrated in **Figure 3**, *p53* follows various mechanisms in distinct cancer types.

Acknowledgements

Merve Nur AL, Berçem Yeman, and Kezban Ucar Ciftci acknowledge YOK 100/2000 Scholarship. Nazlican Yurekli acknowledges TUBITAK 2247-C Scholarship.

Author details

Kubra Acikalin Coskun[1], Merve Tutar[2], Mervenur Al[3], Asiye Gok Yurttas[4],
Elif Cansu Abay[5], Nazlican Yurekli[5], Bercem Yeman Kiyak[6], Kezban Ucar Cifci[6]
and Yusuf Tutar[7,8,9]*

1 Faculty of Medicine, Division of Medicinal Biology, Department of Basic Sciences, Istanbul Aydin University, Istanbul, Turkey

2 Department of Molecular Biology and Genetics, Bilkent University, Ankara, Turkey

3 Faculty of Medicine, Division of Medicinal Biochemistry, Department of Basic Sciences, Hamidiye University of Health Sciences-Turkey, Istanbul, Turkey

4 Faculty of Pharmacy, Division of Biochemistry, Department of Basic Pharmaceutical Sciences, Istanbul Health and Technology University, Istanbul, Turkey

5 Hamidiye Faculty of Medicine, Division of Medicinal Biology, Department of Basic Sciences, University of Health Sciences-Turkey, Istanbul, Turkey

6 Division of Molecular Medicine, Hamidiye Health Sciences Institutes, University of Health Sciences-Turkey, Istanbul, Turkey

7 Hamidiye Faculty of Pharmacy, Division of Biochemistry, Department of Basic Pharmaceutical Sciences, University of Health Sciences-Turkey, Istanbul, Turkey

8 Division of Molecular Oncology, Hamidiye Health Sciences Institutes, University of Health Sciences-Turkey, Istanbul, Turkey

9 Validebağ Research Center, University of Health Sciences-Turkey, Istanbul, Turkey

*Address all correspondence to: yusuf.tutar@sbu.edu.tr

IntechOpen

References

[1] Xu Z, Wu W, Yan H, Hu Y, He Q, Luo P. Regulation of p53 stability as a therapeutic strategy for cancer. Biochemical Pharmacology. 2021; **185**:114407. DOI: 10.1016/j.bcp.2021. 114407. Epub 2021 Jan 7

[2] Hoffman R, Benz EJ, Silberstein LE, Heslop H, Weitz JI, Anastasi J, et al. Hematology: Basic principles and practice. 2018. Available from: https://nls.ldls.org.uk/welcome. html?ark:/81055/vdc_100047773037. 0x000001

[3] Demir Ö, Barros EP, Offutt TL, Rosenfeld M, Amaro RE. An integrated view of p53 dynamics, function, and reactivation. Current Opinion in Structural Biology. 2021;**67**:187-194. DOI: 10.1016/j.sbi.2020.11.005. Epub 2021 Jan 2

[4] Giacinti C, Giordano A. RB and cell cycle progression. Oncogene. 2006;**25**(38):5220-5227. DOI: 10.1038/ sj.onc.1209615

[5] Kastan MB, Bartek J. Cell-cycle checkpoints and cancer. Nature. 2004;**432**(7015):316-323. DOI: 10.1038/ nature03097

[6] Bargonetti J, Prives C. Gain-of-function mutant p53: History and speculation. Journal of Molecular Cell Biology. 2019;**11**(7):605-609. DOI: 10.1093/jmcb/mjz067

[7] Carroll JL, Michael Mathis J, Bell MC, Santoso JT. p53 adenovirus as gene therapy for ovarian cancer. Methods in Molecular Medicine. 2001;**39**:783-792. DOI: 10.1385/1-59259-071-3:783

[8] Lacroix M, Riscal R, Arena G, Linares LK, Le Cam L. Metabolic functions of the tumor suppressor p53: Implications in normal physiology, metabolic disorders, and cancer. Molecular Metabolism. 2020;**33**:2-22.

DOI: 10.1016/j.molmet.2019.10.002. Epub 2019 Oct 18

[9] Yamasaki L. Role of the RB tumor suppressor in cancer. Cancer Treatment and Research. 2003;**115**:209-239. DOI: 10.1007/0-306-48158-8_9

[10] Hu J, Cao J, Topatana W, Juengpanich S, Li S, Zhang B, et al. Targeting mutant p53 for cancer therapy: Direct and indirect strategies. Journal of Hematology & Oncology. 2021;**14**(1):157. DOI: 10.1186/ s13045-021-01169-0

[11] Biasoli D, Kahn SA, Cornélio TA, Furtado M, Campanati L, Chneiweiss H, et al. Retinoblastoma protein regulates the crosstalk between autophagy and apoptosis, and favors glioblastoma resistance to etoposide. Cell Death & Disease. 2013;**4**(8):e767. DOI: 10.1038/ cddis.2013.283

[12] Zhang T, Hu H, Yan G, Wu T, Liu S, Chen W, et al. Long non-coding RNA and breast cancer. Technology in Cancer Research & Treatment. 2019;**18**:1533033819843889. DOI: 10.1177/1533033819843889

[13] Bertheau P, Lehmann-Che J, Varna M, Dumay A, Poirot B, Porcher R, et al. p53 in breast cancer subtypes and new insights into response to chemotherapy. Breast. 2013;**22**(Suppl. 2): S27-S29. DOI: 10.1016/j.breast.2013. 07.005

[14] Bronner SM, Merrick KA, Murray J, Salphati L, Moffat JG, Pang J, et al. Design of a brain-penetrant CDK4/6 inhibitor for glioblastoma. Bioorganic & Medicinal Chemistry Letters. 2019;**29**(16):2294-2301. DOI: 10.1016/j. bmcl.2019.06.021 Epub 2019 Jun 26

[15] Januškevičienė I, Petrikaitė V. Heterogeneity of breast cancer: The importance of interaction between different tumor cell populations.

Life Sciences. 2019;**239**:117009.
DOI: 10.1016/j.lfs.2019.117009. Epub
2019 Oct 24

[16] Dittmer J. Breast cancer stem cells:
Features, key drivers and treatment
options. Seminars in Cancer Biology.
2018;**53**:59-74. DOI: 10.1016/j.semcancer.
2018.07.007. Epub 2018 Jul 27

[17] Kast K, Link T, Friedrich K,
Petzold A, Niedostatek A, Schoffer O,
et al. Impact of breast cancer subtypes
and patterns of metastasis on outcome.
Breast Cancer Research and Treatment.
2015;**150**(3):621-629. DOI: 10.1007/
s10549-015-3341-3. Epub 2015 Mar 18

[18] Bellazzo A, Sicari D, Valentino E,
Del Sal G, Collavin L. Complexes
formed by mutant p53 and their roles in
breast cancer. Breast Cancer. 2018;
10:101-112. DOI: 10.2147/BCTT.S145826

[19] Prabhu KS, Raza A, Karedath T,
Raza SS, Fathima H, Ahmed EI, et al.
Non-coding RNAs as regulators and
markers for targeting of breast cancer
and cancer stem cells. Cancers (Basel).
2020;**12**(2):351. DOI: 10.3390/cancers
12020351

[20] Abdeen SK, Aqeilan RI. Decoding
the link between WWOX and p53 in
aggressive breast cancer. Cell Cycle.
2019;**18**(11):1177-1186. DOI: 10.1080/
15384101.2019.1616998. Epub
2019 May 16

[21] Gasco M, Shami S, Crook T.
The p53 pathway in breast cancer. Breast
Cancer Research. 2002;**4**(2):70-76.
DOI: 10.1186/bcr426. Epub 2002 Feb 12

[22] Lim LY, Vidnovic N, Ellisen LW,
Leong CO. Mutant p53 mediates
survival of breast cancer cells. British
Journal of Cancer. 2009;**101**(9):1606-
1612. DOI: 10.1038/sj.bjc.6605335. Epub
2009 Sep 22

[23] Xiao G, Lundine D, Annor GK,
Canar J, Ellison V, Polotskaia A, et al.
Gain-of-function mutant p53 R273H

Interacts with replicating DNA and
PARP1 in breast cancer. Cancer
Research. 2020;**80**(3):394-405.
DOI: 10.1158/0008-5472.CAN-19-1036

[24] Zhang C, Liu J, Xu D, Zhang T,
Hu W, Feng Z. Gain-of-function mutant
p53 in cancer progression and therapy.
Journal of Molecular Cell Biology.
2020;**12**(9):674-687. DOI: 10.1093/
jmcb/mjaa040

[25] Synnott NC, O'Connell D, Crown J,
Duffy MJ. COTI-2 reactivates mutant
p53 and inhibits growth of triple-
negative breast cancer cells. Breast
Cancer Research and Treatment.
2020;**179**(1):47-56. DOI: 10.1007/
s10549-019-05435-1. Epub 2019 Sep 19

[26] Janani SK, Dhanabal SP,
Sureshkumar R, Surya NUS,
Chenmala K. Guardian of genome on
the tract: Wild type p53-mdm2 complex
inhibition in healing the breast cancer.
Gene. 2021;**786**:145616. DOI: 10.1016/j.
gene.2021.145616. Epub 2021 Apr 1

[27] Srivastava P, Jaiswal PK, Singh V,
Mittal RD. Role of p53 gene
polymorphism and bladder cancer
predisposition in northern India. Cancer
Biomarkers. 2011;**8**:21-28. DOI: 10.3233/
DMA-2011-0816

[28] Hollstein M, Rice K, Greenblatt MS,
Soussi T, Fuchs R, Sørlie T, et al.
Database of p53 gene somatic mutations
in human tumors and cell lines. Nucleic
Acids Research. 1994;**22**(17):3551-3555

[29] DeWolf WC. p53: An important key
to understanding urologic cancer. AUA
Update Series. 1995;**15**:258

[30] Solomon JP, Hansel DE. The
emerging molecular landscape of
urothelial carcinoma. Surgical
Pathology Clinics. 2016;**9**(3):391-404.
DOI: 10.1016/j.path.2016.04.004

[31] Kastenhuber ER, Lowe SW. Putting
p53 in context. Cell. 2017;**170**(6):1062-
1078. DOI: 10.1016/j.cell.2017.08.028

[32] Mansor SF. Manipulation of p53 protein in bladder cancer treatment. IIUM Medical Journal Malaysia. Jan 2021;**20**(1). DOI: 10.31436/imjm. v20i1.1764

[33] Knowles MA, Hurst CD. Molecular biology of bladder cancer: New insights into pathogenesis and clinical diversity. Nature Reviews Cancer. 2015;**15**(1):25

[34] Katkoori V, Jia X, Shanmugam C. Prognostic significance of p53 Codon 72 polymorphism differs with race in colorectal adenocarcinoma. Clinical Cancer Research. 2009;**15**:2406-2416

[35] Berggren P, Steineck G, Adolfsson J, Hansson J, Jansson O, Larsson P, et al. p53 mutations in urinary bladder cancer. British Journal of Cancer. 2001;**84**(11):1505-1511

[36] Thomas M, Kalita A, Labrecque S, Pim D, Banks L, Matlashewski G. Two polymorphic variants of wild-type p53 differ biochemically and biologically. Molecular and Cellular Biology. 1999;**19**(2);1092-1100. DOI: 10.1128/MCB.19.2.1092

[37] Chen W, Tsai F, Wu J, Wu H, Lu H, Li C. Distributions of p53 codon 72 polymorphism in bladder cancer—proline form is prominent in invasive tumor. Urological Research. 2000;**28**(5):293-296

[38] Soulitzis N, Sourvinos G, Dokianakis D, Spandidos D. p53 codon 72 polymorphism and its association with bladder cancer. Cancer Letters. 2002;**179**:175-183

[39] Kuroda Y, Tsukino H, Nakao H, Imai H, Katoh T. p53 codon 72 polymorphism and urothelial cancer risk. Cancer Letters. 2003;**189**:77-83

[40] Whibley C, Pharoah P, Hollstein M. p53 polymorphisms: Cancer implications. Nature Reviews Cancer. 2009;**9**:95-107

[41] Schulz W. Molecular Biology of Human Cancers. Netherlands: Springer; 2007

[42] Lapointe S, Perry A, Butowski NA. Primary brain tumours in adults. Lancet. 2018;**392**(10145):432-446. DOI: 10.1016/S0140-6736(18)30990-5. Epub 2018 Jul 27

[43] Vargo MM. Brain tumors and metastases. Physical Medicine and Rehabilitation Clinics of North America. 2017;**28**(1):115-141. DOI: 10.1016/j.pmr.2016.08.005

[44] Checler F, Alves da Costa C. p53 in neurodegenerative diseases and brain cancers. Pharmacology & Therapeutics. 2014;**142**(1):99-113. DOI: 10.1016/j.pharmthera.2013.11.009. Epub 2013 Nov 25

[45] Ham SW, Jeon HY, Jin X, Kim EJ, Kim JK, Shin YJ, et al. TP53 gain-of-function mutation promotes inflammation in glioblastoma. Cell Death and Differentiation. 2019;**26**(3):409-425. DOI: 10.1038/s41418-018-0126-3. Epub 2018 May 21

[46] Lah TT, Novak M, Breznik B. Brain malignancies: Glioblastoma and brain metastases. Seminars in Cancer Biology. 2020;**60**:262-273. DOI: 10.1016/j.semcancer.2019.10.010. Epub 2019 Oct 22

[47] Agostini M, Melino G, Bernassola F. The p53 family in brain disease. Antioxidants & Redox Signaling. 2018;**29**(1):1-14. DOI: 10.1089/ars.2017.7302. Epub 2017 Nov 27

[48] Hooper C, Meimaridou E, Tavassoli M, Melino G, Lovestone S, Killick R. p53 is upregulated in Alzheimer's disease and induces tau phosphorylation in HEK293a cells. Neuroscience Letters. 2007;**418**(1):34-37. DOI: 10.1016/j.neulet.2007.03.026. Epub 2007 Mar 15

[49] LaFerla FM, Hall CK, Ngo L, Jay G. Extracellular deposition of beta-amyloid upon p53-dependent neuronal cell death in transgenic mice. The Journal of Clinical Investigation. 1996;**98**(7):1626-1632. DOI: 10.1172/JCI118957

[50] Xiong Y, Zhang Y, Xiong S, Williams-Villalobo AE. A glance of p53 functions in brain development, neural stem cells, and brain cancer. Biology (Basel). 2020;**9**(9):285. DOI: 10.3390/biology9090285

[51] Armstrong JF, Kaufman MH, Harrison DJ, Clarke AR. High-frequency developmental abnormalities in p53-deficient mice. Current Biology. 1995;**5**(8):931-936. DOI: 10.1016/s0960-9822(95)00183-7

[52] Pedrote MM, Motta MF, Ferretti GDS, Norberto DR, Spohr TCLS, Lima FRS, et al. Oncogenic gain of function in glioblastoma is linked to mutant p53 amyloid oligomers. iScience. 2020;**23**(2):100820. DOI: 10.1016/j.isci.2020.100820. Epub 2020 Jan 8

[53] Guo CF, Zhuang Y, Chen Y, Chen S, Peng H, Zhou S. Significance of tumor protein p53 mutation in cellular process and drug selection in brain lower grade (WHO grades II and III) glioma. Biomarkers in Medicine. 2020;**14**(12):1139-1150. DOI: 10.2217/bmm-2020-0331 Epub 2020 Jul 15

[54] Cao H, Chen X, Wang Z, Wang L, Xia Q, Zhang W. The role of MDM2-p53 axis dysfunction in the hepatocellular carcinoma transformation. Cell Death Discovery. 2020;**6**:53. DOI: 10.1038/s41420-020-0287-y

[55] Fang SS, Guo JC, Zhang JH, Liu JN, Hong S, Yu B, et al. A P53-related microRNA model for predicting the prognosis of hepatocellular carcinoma patients. Journal of Cellular Physiology. 2020;**235**(4):3569-3578. DOI: 10.1002/jcp.29245. Epub 2019 Sep 25

[56] Caruso S, O'Brien DR, Cleary SP, Roberts LR, Zucman-Rossi J. Genetics of hepatocellular carcinoma: Approaches to explore molecular diversity. Hepatology. 2021;**73**(Suppl. 1):14-26. DOI: 10.1002/hep.31394. Epub 2020 Dec 8

[57] Ozaki T, Nakagawara A. Role of p53 in cell death and human cancers. Cancers (Basel). 2011;**3**(1):994-1013. DOI: 10.3390/cancers3010994

[58] Meng X, Franklin DA, Dong J, Zhang Y. MDM2-p53 pathway in hepatocellular carcinoma. Cancer Research. 2014;**74**(24):7161-7167. DOI: 10.1158/0008-5472.CAN-14-1446. Epub 2014 Dec 4

[59] Hussain SP, Schwank J, Staib F, Wang XW, Harris CC. TP53 mutations and hepatocellular carcinoma: Insights into the etiology and pathogenesis of liver cancer. Oncogene. 2007;**26**(15):2166-2176. DOI: 10.1038/sj.onc.1210279

[60] Kunst C, Haderer M, Heckel S, Schlosser S, Mueller M. The p53 family in hepatocellular carcinoma. Translational Cancer Research;**5**(6):632-638. DOI: 10.21037/tcr.2016.11.79

[61] Yang C, Huang X, Li Y, Chen J, Lv Y, Dai S. Prognosis and personalized treatment prediction in TP53-mutant hepatocellular carcinoma: An in silico strategy towards precision oncology. Briefings in Bioinformatics. 2021;**22**(3):bbaa164. DOI: 10.1093/bib/bbaa164

[62] Ottaviani G, Jaffe N. The epidemiology of osteosarcoma. Cancer Treatment and Research. 2009;**152**:3-13. DOI: 10.1007/978-1-4419-0284-9_1

[63] Liu P, Wang M, Li L, Jin T. Correlation between osteosarcoma and the expression of WWOX and p53. Oncology Letters. 2017;**14**(4):4779-4783. DOI: 10.3892/ol.2017.6747. Epub 2017 Aug 10

[64] Overholtzer M, Rao PH, Favis R, Lu XY, Elowitz MB, Barany F, et al. The presence of p53 mutations in human osteosarcomas correlates with high levels of genomic instability. Proceedings of the National Academy of Sciences of the United States of America. 2003;**100**(20):11547-11552. DOI: 10.1073/pnas.1934852100. Epub 2003 Sep 12. Erratum in: Proceedings of the National Academy of Sciences of the United States of America. 2003;**100**(24): 14511

[65] Chen Z, Guo J, Zhang K, Guo Y. TP53 mutations and survival in osteosarcoma patients: A meta-analysis of published data. Disease Markers. 2016;**2016**:4639575. DOI: 10.1155/2016/4639575. Epub 2016 Apr 27

[66] Chen X, Bahrami A, Pappo A. Recurrent somatic structural variations contribute to tumorigenesis in pediatric osteosarcoma. Cell Reports. 2014;**7**(1): 104-112. DOI: 10.1016/j.celrep.2014.03.003

[67] Mogi A, Kuwano H. TP53 mutations in nonsmall cell lung cancer. Journal of Biomedicine & Biotechnology. 2011;**2011**:583929. DOI: 10.1155/2011/583929. Epub 2011 Jan 18

[68] Steels E, Paesmans M, Berghmans T, Branle F, Lemaitre F, Mascaux C, et al. Role of p53 as a prognostic factor for survival in lung cancer: A systematic review of the literature with a meta-analysis. The European Respiratory Journal. 2001;**18**(4):705-719. DOI: 10.1183/09031936.01.00062201

[69] Sameshima Y, Matsuno Y, Hirohashi S, Shimosato Y, Mizoguchi H, Sugimura T, et al. Alterations of the p53 gene are common and critical events for the maintenance of malignant phenotypes in small-cell lung carcinoma. Oncogene. 1992;**7**(3): 451-457

[70] Tammemagi MC, McLaughlin JR, Bull SB. Meta-analyses of p53 tumor suppressor gene alterations and clinicopathological features in resected lung cancers. Cancer Epidemiology, Biomarkers & Prevention. 1999;**8**(7): 625-634

[71] Chang YL, Wu CT, Shih JY, Lee YC. Comparison of p53 and epidermal growth factor receptor gene status between primary tumors and lymph node metastases in non-small cell lung cancers. Annals of Surgical Oncology. 2011;**18**(2):543-550. DOI: 10.1245/s10434-010-1295-6. Epub 2010 Sep 2

[72] Parkin DM, Bray F, Ferlay J, Pisani P. Global cancer statistics, 2002. CA: A Cancer Journal for Clinicians. 2005;**55**(2):74-108. DOI: 10.3322/canjclin.55.2.74

[73] Hainaut P, Pfeifer GP. Patterns of p53 G→T transversions in lung cancers reflect the primary mutagenic signature of DNA-damage by tobacco smoke. Carcinogenesis. 2001;**22**(3):367-374. DOI: 10.1093/carcin/22.3.367

[74] Pfeifer GP, Denissenko MF, Olivier M, Tretyakova N, Hecht SS, Hainaut P. Tobacco smoke carcinogens, DNA damage and p53 mutations in smoking-associated cancers. Oncogene. 2002;**21**(48):7435-7451. DOI: 10.1038/sj.onc.1205803

[75] Denissenko MF, Pao A, Tang M, Pfeifer GP. Preferential formation of benzo[a]pyrene adducts at lung cancer mutational hotspots in P53. Science. 1996;**274**(5286):430-432. DOI: 10.1126/science.274.5286.430

[76] Smith LE, Denissenko MF, Bennett WP, Li H, Amin S, Tang M, et al. Targeting of lung cancer mutational hotspots by polycyclic aromatic hydrocarbons. Journal of the National Cancer Institute. 2000;**92**(10):803-811. DOI: 10.1093/jnci/92.10.803

[77] Kondo K, Tsuzuki H, Sasa M, Sumitomo M, Uyama T, Monden Y. A

dose-response relationship between the frequency of p53 mutations and tobacco consumption in lung cancer patients. Journal of Surgical Oncology. 1996;**61**(1):20-26. DOI: 10.1002/(SICI)1096-9098(199601)61:1<20::AID-JSO6>3.0.CO;2-U

[78] Gow CH, Chang YL, Hsu YC, Tsai MF, Wu CT, Yu CJ, et al. Comparison of epidermal growth factor receptor mutations between primary and corresponding metastatic tumors in tyrosine kinase inhibitor-naive non-small-cell lung cancer. Annals of Oncology. 2009;**20**(4):696-702. DOI: 10.1093/annonc/mdn679. Epub 2008 Dec 16

[79] Vogt U, Zaczek A, Klinke F, Granetzny A, Bielawski K, Falkiewicz B. p53 Gene status in relation to ex vivo chemosensitivity of non-small cell lung cancer. Journal of Cancer Research and Clinical Oncology. 2002;**128**(3):141-147. DOI: 10.1007/s00432-001-0305-2. Epub 2002 Jan 26

[80] Leslie WT, Bonomi PD. Novel treatments in non-small cell lung cancer. Hematology/Oncology Clinics of North America. 2004;**18**(1):245-267. DOI: 10.1016/s0889-8588(03)00146-1

[81] Swisher SG, Roth JA, Nemunaitis J. Adenovirus-mediated p53 gene transfer in advanced non-small-cell lung cancer. Journal of the National Cancer Institute. 1999;**91**(9):763-771. DOI: 10.1093/jnci/91.9.763

[82] Issaeva N, Bozko P, Enge M, Protopopova M, Verhoef LG, Masucci M, et al. Small molecule RITA binds to p53, blocks p53-HDM-2 interaction and activates p53 function in tumors. Nature Medicine. 2004;**10**(12):1321-1328. DOI: 10.1038/nm1146. Epub 2004 Nov 21

[83] Selivanova G, Wiman KG. Reactivation of mutant p53: Molecular mechanisms and therapeutic potential. Oncogene. 2007;**26**(15):2243-2254. DOI: 10.1038/sj.onc.1210295

[84] Magrini R, Russo D, Ottaggio L, Fronza G, Inga A, Menichini P. PRIMA-1 synergizes with adriamycin to induce cell death in non-small cell lung cancer cells. Journal of Cellular Biochemistry. 2008;**104**(6):2363-2373. DOI: 10.1002/jcb.21794

Chapter 3

The Role p53 Protein in DNA Repair

Bakhanashvili Mary

Abstract

The tumor suppressor p53 protein controls cell cycle and plays a vital role in preserving DNA integrity. p53 is activated by varied stress signals and the distribution of p53 between the different subcellular compartments depends on the cellular stress milieu. DNA repair pathways protect cells from damage that can lead to DNA breaks. The multi-functional p53 protein promotes DNA repair both directly and indirectly through multiple mechanisms; it accomplishes multi-compartmental functions by either numerous p53-controlled proteins or by its inherent biochemical activities. Accumulating evidence supports the contribution of p53 in the maintenance of the genomic integrity and in various steps of the DNA damage response, through its translocation into nucleus and mitochondria. p53 may also be utilized by viral polymerases in cytoplasm to maintain genomic integrity of viruses, thus expanding the role of p53 as a 'guardian of the genome'. We summarize recent findings highlighting roles of p53 in DNA repair.

Keywords: p53, DNA repair, mitochondrial DNA, viral DNA

1. Introduction

Humans are persistently exposed to various chemical and physical agents that have the potential to damage genomic DNA, such as, irradiation (IR), ultraviolet (UV) light, reactive oxygen species (ROS), et cetera [1]. The integrity and survival of a cell is critically dependent on genome stability and mammalian cells have established multiple pathways to repair different types of target DNA lesions to safeguard the genome from deleterious consequences of various kinds of stresses [2]. The significance of the DNA repair in the protection of genomic stability is highlighted by the fact that many proteins/factors involved have been preserved through evolution [3].

DNA damage, induced by endogenous and exogenous agents, is a common event and must undergo a variety of DNA damage repair in order to ensure the faithful transfer of genetic information during cell division [3]. Four main DNA polymerases are involved with nuclear DNA replication: DNA polymerase α, β, δ and ε [1] (**Figure 1**). DNA repair pathways, which are also recognized as guardians of the genome, protect cells from numerous damages leading to DNA breaks [4]. Failure to restore DNA lesions or inappropriate repair of DNA damage give rise to genomic instability, which is a hallmark of cancer. Remarkably, mild and massive DNA damage are differentially integrated into the cellular signaling networks and, in consequence, provoke different cell fate decisions. After mild damage, the cellular response is cell cycle arrest, DNA repair, and cell survival, whereas severe damage,

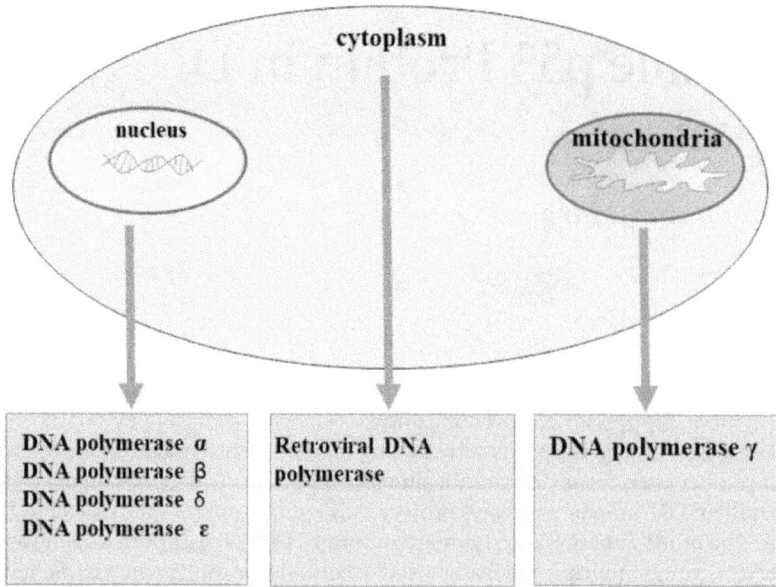

Figure 1.
Sub-cellular localization of eukaryotic and retroviral DNA polymerases.

drives the cell death response. The inability of the DNA damage response (DDR) to repair following endogenous and exogenous insults can lead to (i) an accumulation of errors in genomic DNA, (ii) subsequent malignant transformation, (iii) cancer progression and (iv) further impairment of the DNA repair capacity. DNA repair mechanisms comprise the detection and deletion (excision) of the lesion, the rejoining of DNA ends and the restoration of the complementary sequence based on a DNA template.

Since cancer cells typically have many mutations compared to a non-cancer cell, it was proposed that one of the earliest changes in the development of a cancer cell is a mutation that increases the spontaneous mutation rate [5]. The presence of a "mutator phenotype" could increase the acquisition of alterations that could lead to enhanced drug resistance limiting the effectiveness of anti-cancer drug treatment.

Viral infection is characterized by the high genetic variability found in virus populations [6]. This phenomenon is attributed to the inaccuracy of the replication machinery that is unique to the viral life cycle. Virulence, pathogenesis and the ability to develop effective antiretroviral drugs and vaccines are largely dependent on genetic diversity in viruses [7]. Retroviruses are RNA viruses that replicate through a DNA intermediate in a process catalyzed by the viral reverse transcriptase (RT) in cytoplasm (**Figure 1**) [7]. Human immunodeficiency virus type 1 (HIV-1), the etiological agent of AIDS, exhibits exceptionally high mutation frequencies [8]. The accepted explanations for the inaccuracy of HIV-1 RT are the relatively low fidelity of the enzyme during DNA synthesis and the deficiency of intrinsic proofreading activity. A strong mutator phenotype is also observed for herpes viral DNA polymerase mutants with reduced intrinsic $3' \to 5'$ exonuclease activity [9].

Mitochondrial DNA (mtDNA) alterations have been associated with various human diseases with impaired mitochondrial function [10]. Mitochondrial DNA polymerase γ (pol γ) is responsible for replication of mtDNA and is implicated in all repair processes (**Figure 1**) [11]. Mitochondrial DNA is prone to mutations, since it is localized near the inner mitochondrial membrane in which reactive oxygen species are generated. Additionally, mtDNA lacks histone protection and the highly efficient DNA repair mechanisms [12]. The mutation rate of mtDNA is estimated to be about 20–100-fold higher than that of nuclear DNA [13]. The mutagenic mechanisms were shown to be replication errors caused by mis insertion (as a result of a dNTP excess), or decreased proofreading efficiency [14, 15].

Thus, in various compartments of the cell, enhanced DNA replication fidelity is a vital activity for the preservation of genomic stability for many organisms.

2. DNA repair

Genomic integrity of the cell is crucial for the successful transmission of genetic information to the offspring and its survival [16]. DNA is constantly being damaged. Essentially, DNA lesions can occur in two major ways, affecting either a single-stranded break (SSB) or double-stranded (DSB) or mono-adducts and inter-strand crosslinks, respectively. To combat this, eukaryotes have developed complex DNA damage repair (DDR) pathways (**Figure 2**). The active pathways for DNA repair are base excision repair (BER), nucleotide excision repair (NER), and mismatch repair MMR for SSB repair, whereas homologous recombination (HR) and non-homologous end-joining (NHEJ) for DSB repair [16]. Nucleotide excision repair (NER) removes a variety of helix-distorting lesions such as typically induced by UV irradiation, whereas base excision repair (BER) targets oxidative base modifications. Mismatch repair (MMR) scans for nucleotides that have been erroneously inserted during replication. The most deleterious types of damage in DNA are DSBs that are typically induced by IR and resolved either by NHEJ or by HR, whereas RECQ helicases assume various roles in genome maintenance during recombination repair and replication.

A low fidelity of DNA synthesis in various compartments of the cell by main replicative DNA polymerases leads to genomic instability (mutator phenotype) [17]. The errors produced during DNA synthesis could result from three fidelity

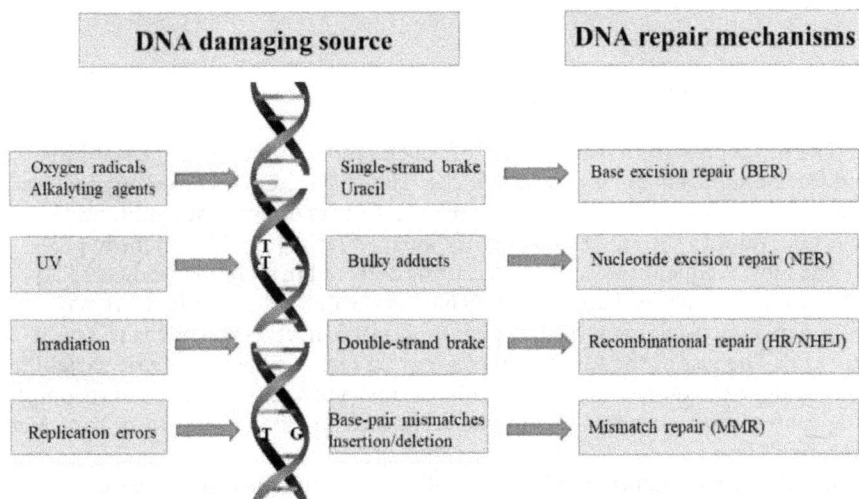

Figure 2.
DNA damage and repair mechanisms. Various DNA damaging agents cause a range of DNA lesions with different outcomes at both the genomic and cellular levels. Each are corrected by a specific DNA repair mechanism, namely, base-excision repair (BER), nucleotide excision repair (NER), homologous recombination (HR)/non-homologous end-joining (NHEJ) or mismatch repair (MMR).

Biochemical properties of cellular DNA polymerases			
	Function	$3' \rightarrow 5'$ exonuclease	Proofreading
Nuclear DNA polymerases			
α	primase	no	no
β	repair	no	no
δ	Lagging DNA synthesis, repair	yes	yes
ε	Leading DNA synthesis, repair	yes	yes
Mitochondrial DNA polymerase			
γ	DNA synthesis	yes	yes
Retroviral DNA polymerase			
HIV-1 RT	DNA synthesis	no	no

Table 1.
Biochemical properties of eukaryotic and retroviral DNA polymerases.

determining processes: a) nucleotide misinsertion into the nascent DNA, b) lack of exonucleolytic proofreading activity, that is, the mechanism to identify and excise incorrect nucleotide incorporated during DNA synthesis, and c) extension of mismatched 3'-termini of DNA (**Table 1**) [18].

Incorrectly repaired DNA lesions can lead to mutations, genomic instability, changes in the regulation of cellular functions, progression of cancer and premature aging. Cells can repair the large variety of DNA lesions through a variety of sophisticated DNA-repair machineries, recognizing and activating battery of proteins/factors for the repair of damaged DNA. DNA replication is a complex process influenced by numerous proteins/factors. The most important part of the DNA damage response is the activation of tumor repressor p53 protein [18].

3. Tumor suppressor p53 protein and DNA repair

The p53 represents a major factor for the maintenance of genome stability and for the suppression of cancer [19, 20]. The p53 protein is commonly referred to as the *"guardian of the genome"* due to its activities directed at maintaining genomic stability through the repair of damaged DNA [19]. Mutations in p53 are the most frequent molecular alterations detected in all human cancers [21]. Approximately 50% of human tumors harbor p53mutations while the remaining malignancies expressing wtp53 display functional inactivation of the p53 pathway [22]. The loss of the functional p53 may be responsible for genetic variability and the development of cancer [22]. Mutations in p53 result in a loss of its physiological function, accompanied by the accumulation of a novel gain-of function protein [23].

Under normal conditions within the cell, p53 is maintained at low levels by the E3 Ubiquitin ligase MDM2, mediating p53 proteasomal degradation [23]. In response to exposure to various endogenous and exogenous stress signals (such as DNA damage, oncogene activation, hypoxia, and nutrient depletion), the protein is stabilized and functionally activated by a series of post-translational modifications (*e.g.*, phosphorylation, acetylation) resulting in p53 accumulation at nuclear and extra-nuclear sites [21, 24]. Activated p53 is a pleiotropic regulator and, as a transcription factor, binds to specific DNA sequences thereby regulating the expression of plethora of target genes controlling proliferation, senescence, DNA repair, and cell death. p53 is involved in diverse cellular processes including cell cycle arrest (thus preventing the replication of damaged DNA allowing time for the cells to repair DNA), apoptosis (for eliminating cells that contained excessive and irreparable damaged DNA), or DNA-damage repair (**Figure 3**) [20, 23]. These processes together protect the organism from genetically unstable cells that drive cancer.

p53 exhibits the functional heterogeneity in its basal (non-induced) state and under various p53 inducible circumstances [20]. Increasing evidences suggest various "non-transcriptional functions" of p53, that can contribute to tumor suppressor activity [25]. p53 may modulate DNA repair through processes, which are independent of its transactivation function. p53 is actively transported between the nucleus and cytoplasm. Furthermore, p53 translocate to mitochondria [26]. p53 can directly interact with DNA repair related cellular factors [27]. The origin, duration, intensity of the stress signals, the interaction with other cellular or viral proteins, and stress-mediated subcellular localization of p53 determines the outcome of the p53 response, namely, its pro- or anti-survival functions [28]. p53 protein executes multi-compartmental functions in the cell by either numerous p53-regulated proteins or by its intrinsic biochemical activities [28].

3.1 p53 and DNA repair in nucleus

The functioning of the eukaryotic genome relies on effective and accurate DNA replication and repair [2]. DNA replication in the nucleus of eukaryotic cells employs DNA polymerases (pols) α, β, δ, and ε, that are the key enzymes required to maintain the integrity of the genome under all these circumstances [1, 3]. However, the maintenance of genomic integrity is complicated by the fact that the genome is persistently challenged by a variety of endogenous and exogenous DNA-damaging factors [4]. DNA lesion can block DNA replication, which can lead to double-strand breaks (DSB) or alter base coding potential, leading to mutations. The accumulation of damage in DNA can affect gene expression leading to the malfunction of many cellular processes [4]. Various DNA repair systems operate in cells to remove DNA lesions, and several proteins are known to be the key components of these repair systems.

The presence of p53 was demonstrated in different nuclear compartments and suggested that the p53 population not engaged in transcriptional regulation could exert functions other than induction of growth arrest or apoptosis and directly participate in processes of repair [25]. p53 mediating various activities are correlated with the levels of the p53 protein in the cells [27, 29]. The non-genotoxic stress may include a long-lasting, moderate accumulation of p53 in nucleus. Conversely, acute genotoxic stress may induce rapid and transient accumulation of very high levels of p53 with preferential activation of target genes involved in apoptosis [29]. There is a possibility that both transcriptional and transcription-independent pathways act in synergy thereby amplifying the potency of involvement of p53 in DNA repair.

p53 localized in cell nuclei in response to replication stress actively participate in various processes of DNA repair and DNA recombination via its ability to interact with components of the repair and recombination machinery and by its various biochemical activities [30, 31]. Both *in vitro* and *in vivo* data suggest an intricate relationship between the biochemical activities of p53 in DNA replication and recombination. The notion that p53 plays a role in DNA repair pathways *in vivo* is supported by the observation that p53 knockout mice exhibit an increase in chromosomal abnormalities and a deficiency in global genomic repair [32]. p53 is

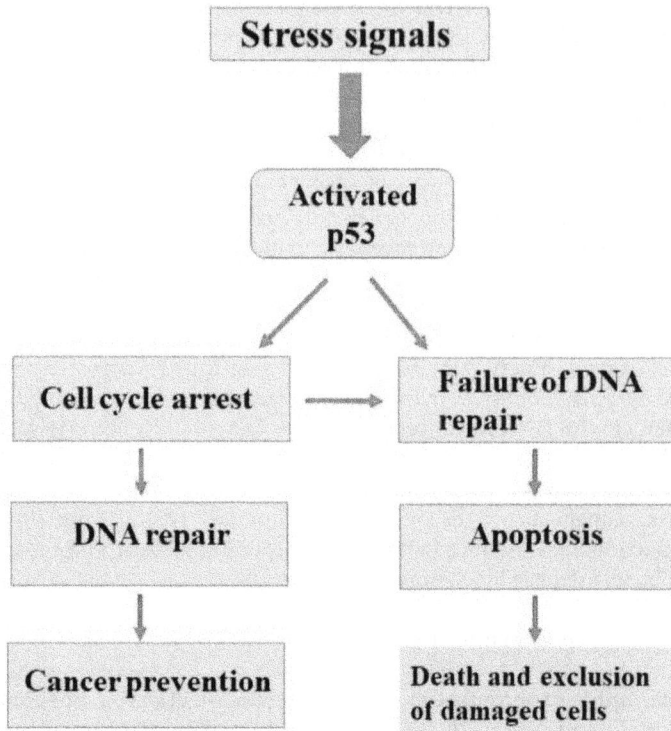

Figure 3.
In response to various endogenous and exogenous stress signals, the activated p53 arrests the cell cycle until the DNA damage is repaired thereby preventing the cancer. If the DNA damage cannot be repaired apoptosis occurs for eliminating cells that contained excessive and irreparable damaged DNA.

involved in almost all nuclear DNA repair pathways including BER, NER, MMR, NHEJ and HR [32]. The transcription-independent functions play a prominent role as a facilitator of DNA repair by halting the cell cycle to allow time for the repair machineries to restore genome stability [25].

The C-terminal 30 amino acids of p53 were shown to recognize several DNA damage-related structures.

In addition, full range of various intrinsic biochemical features of the p53 protein support its possible roles in DNA repair. After DNA damage: (a) p53 is able to recognize and bind sites of DNA damage, such as ssDNA and dsDNA ends [33, 34], (b) p53 catalyzes DNA and RNA strand transfer and promotes the annealing of complementary DNA and RNA single-strands [35, 36], (c) p53 binds insertion/deletion mismatches and bulges [37], (d) p53 binds to three-stranded heteroduplex joints and four-stranded Holliday junction DNA structures with localization specifically at the junction, suggesting that p53 directly participates in recombination repair [38], (e) it can bind DNA in a non-sequence-specific manner [39], (f) p53 exhibits a Mg2+ dependent $3' \rightarrow 5'$ exonuclease activity [40–43].

Noticeably, the same central region within p53, where tumorigenic mutations are clustered, recognizes DNA sequence specifically, is required for junction-specific binding of heteroduplex joints and is necessary and sufficient for the $3' \rightarrow 5'$ exonuclease activity on DNA [28]. In addition to p53's biochemical activities, numerous reports on physical and functional protein interactions further strengthened the proposal of a direct role of p53 in BER, NER, and DSB repair.

a. Oxidative DNA damage is largely repaired by the BER pathway. p53 might directly facilitate BER mainly via association with BER components. Wtp53 directly enhanced BER activity measured both *in vitro* and *in vivo* [44]. Genotoxic stress induced a p53-dependent modulation in BER activity throughout the cell cycle. The idea that p53 is directly involved in BER is supported by various studies: BER activity in cell extracts correlates with levels of purified wtp53 [29], the ability of p53 to augment BER activity is correlated with its ability to interact directly both with AP endonuclease and with DNA Polymerase β [27, 45]. Hot-spot tumor-derived p53 mutants do not significantly enhance BER, supporting the possibility that the stimulatory effect of wtp53 may contribute to its ability to suppress tumorigenesis. Based on these results, p53 stabilization of the DNA pol β–AP-DNA complex is likely to be the mechanism underlying the stimulation of BER by p53 [27].

The cellular response depends on the dose of genotoxic agent introduced to the cells. Increasing doses of genotoxic agents cause the accumulation of activated p53 that determines the onset of BER or apoptosis. Low doses of DNA damaging agent resulted in the enhancement of p53-dependent BER activity whereas high levels induced different p53 post-translational modifications that down regulate BER pathway and instead provoked an apoptotic response [29]. The quantitative changes in p53 protein level were associated with qualitative changes in p53 phosphorylation status. In all, this may indicate that increasing doses of genotoxic agents cause the accumulation of activated p53 that determines the onset of BER or apoptosis.

b. NER is an important DNA repair process that detects and eliminates lesions including both chemical alteration and structural distortion of the DNA helix (*e.g.*, photoproducts induced by UV irradiation and other bulky lesions) [25]. The NER pathway retains two damage detection pathways: Transcription Coupled Repair (TC-NER) and Global Genome Repair (GG-NER), depending on the mode of damage recognition in the entire genome versus actively transcribed regions [46–48]. TC-NER detects and removes transcription blocking lesions in transcribed sections of the genome; triggered when a lesion inhibits transcription elongation by RNA polymerase II, thereby preventing cell death. GGR-NER detects lesions across the whole genome, including non-transcribed regions. Upon lesion detection by either the TC or GG arm, repair proceeds via a final common pathway [25]. The role of p53 in promoting GG-NER is more consistent compared to p53 function in TC-NER. p53 facilitates NER by promoting lesion recognition or detection by recruiting the p300 histone acetylase to damage sites, which acetylates the histone H3, leading to global chromatin relaxation and increased lesion accessibility making an additional contribution to the maintenance of genome stability [46–47].

Pathogenic mutations in the GG components XPC and DDB2 (XPE) result in xeroderma pigmentosum (XP) a disease characterized by increased UV-sensitivity and skin cancer incidence [46]. Conversely, mutation in TC genes result in Cockayne's syndrome that is characterized by neurological abnormalities but no increase in skin cancer incidence. Some NER proteins, particularly the GG damage recognition proteins, can decide a cell's fate by triggering the initiation of the repair pathway or by signaling apoptosis [46]. Therefore, if the GG pathway is defective, neither DNA repair nor apoptosis occurs, resulting in a cancer cell containing high levels of UV-induced mutations that does not undergo apoptosis. How this non-transcriptional function of p53 contributes to tumor suppression is unclear.

c. DNA mismatch repair (MMR) is an important DNA repair pathway, which facilitates removal of incorrect nucleotides incorporated during replication. p53 facilitates excision of incorrect nucleotides produced from the error prone nature of DNA polymerases and misincorporation of the incorrect base [25]. Mismatched bases can be either a G/T or A/C pair. To initiate MMR a nick in the DNA either 5' or 3' to the mismatch must occur. Proteins that bind the mismatch in humans are *E. coli* MutS homologs. MSH2 is a major component of the MMR MSH2-MSH6 complex and is a known to be transcriptionally upregulated by p53 following UV [49]. In vitro studies established that the MSH2–MSH6 complex can enhance the binding of p53 to DNA substrates with topological distortions, and this activity depends on the phosphorylation state of p53(S392) [50, 51]. Connections between p53 and MMR have been made in various systems demonstrating a role for MMR proteins in influencing p53-related processes. p53 and MMR proteins can function synergistically in mice, as Msh2−/− p53−/− females arrested as embryos and they quickly developed tumors relative to the single-mutant animals [52]. p53 signaling was shown to be suppressed in MSH2-deficient cells [53]. While these p53-dependent mechanisms have been linked to MMR regulation, MSH2 has been implicated in a variety of repair pathways and it is necessary to determine if p53 function is pertinent. An interesting notion is that, p53 interacts with and transcription-ally regulates its gene target in MMR. Further studies are needed to define if p53 transcription-dependent and independent functions work alongside in MMR or whether these functions are separate and dependent on the cellular insult or pathway choice.

d. Mutator phenotypes (with the potential for cancer progression) have been reported for cells that lack a proofreading 3' → 5' exonuclease activity associated with the DNA polymerase [54]. Excision of incorrectly polymerized nucleotides by exonucleases is an imperious mechanism diminishing the errors during DNA polymerization [55]. Certain organisms with a deficiency of exonucleolytic proofreading, have an increased susceptibility to cancer, especially under conditions of stress. Because the misincorporation of non-complementary dNTPs during DNA replication represents a chief mechanism of gene mutation [56], the removal of the wrong nucleotides from DNA is critical for genomic stability. The intrinsic limited accuracy of DNA polymerases and the imbalance of intracellular dNTP pools are the two most important factors responsible for DNA replication errors [57, 58]. The proofreading for such replication errors by the 3' → 5' exonuclease activity associated with the DNA replication machinery is extremely important in reduction of the occurrence of mutations. Interestingly, the mammalian DNA pol α, an enzyme considered to be responsible for the lagging strand replication [59], lacks the 3' → 5' exonuclease proof-reading activity and is prone to making replication errors [60].

Three steps, base selection, exonucleolytic proofreading, and DNA elongation, ensure the high fidelity of DNA replication. wtp53 exhibits an intrinsic 3' → 5' exonuclease activity. wtp53, co-located with the DNA replication machinery [61], specifically interacts with pol α and has been shown to preferentially eliminate mismatched nucleotides from DNA with its 3' → 5' exonuclease activity, thereby enhancing the DNA replication fidelity of pol α *in vitro* [41].

Hydroxyurea (HU), an inhibitor of ribonucleotide reductase involved in the *de novo* synthesis of deoxynucleotides, was used to induce dNTP pool imbalance and to cause mutations in the cells due to misincorporation of unpaired deoxynucleotides into DNA [62]. The examination of the rates of HU-induced mutations in H1299 (p53-null) and H460 (wtp53) cells discovered substantially augmented mutation rates in H1299 cells. Furthermore, the HU-induced mutation frequency was significantly reduced by introduction of wtp53 expression vector into the p53-null H1299 cells. Thus, wtp53 expression was associated with a reduction of mutations caused by replication errors under the stress of dNTP pool imbalance [62].

The functional interaction of DNA polymerase and exonuclease activity was observed with p53/pol-prim complex. p53-containing DNA pol-prim complex excised preferentially a 3'-mispaired primer end over a paired one and replaced it with a correctly paired nucleotide [63]. In contrast, a pol-prim complex containing the hot spot mutant p53R248H did not display exonuclease activity and did not elongate a mispaired 3'-end, representing that the p53 exonuclease from the p53/pol-prim complex was indispensable for the subsequent elongation of the primer by DNA polymerase. These findings support the view that p53 might fulfill a proofreading function for pol-prim and suggest that the defect in proofreading function of p53 may contribute to genetic instability associated with cancer development and progression [63].

e. DSBs are the most severe type of DNA damage, and these DSBs generated at the replication fork are repaired by two principal repair pathways: homology-based repair (HR) and non-homologous end-joining (NHEJ) [25, 31]. Furthermore, replication blocking lesions such as bulky adducts are subject to HR repair, thereby rescuing the replication fork. HR is considered the most error-free pathway, because sister chromatids are the preferred template, however, it can also produce genetic instability upon up- or down-regulation [25].

Depending on the type and quality of the DSB repair pathway involved, the repair process may end up with deletions, loss of heterozygosity, and chromosomal translocations which may accelerate the multistep process of tumorigenesis. p53 can control HR *in vitro* by specific recognition of the heteroduplex intermediates, and *in vivo* by modulating the functions of different HR-specific proteins [38, 64]. Numerous groups detected that wtp53 represses HR on both extra-chromosomal and intra-chromosomal DNA substrates by at least one to two orders of magnitude [31]. Conversely, inactivation of p53 by mutation or complex formation by viral proteins increased HR by several orders of magnitude. Importantly, experiments with p53 mutants revealed severe HR inhibitory defects for all tested hotspot mutants. Mutant p53s which are known to reduce or even abolish p53's transcriptional transactivation and cell cycle regulatory capacity, did not significantly affect HR inhibition [65, 66]. These discoveries confirmed that p53 activities in transcriptional transactivation and checkpoint control are separable from its functions in homology-based DSB repair and provided undoubted proof for a direct role of p53 in HR [67].

p53 prevents the accumulation of DSBs at stalled-replication forks induced by UV or hydroxyurea (HU) treatment. When DNA replication is blocked, p53 becomes phosphorylated on serine 15 and associates with key enzymes of HR such as, Rad51, and Rad54 [68, 69]. Notably, during replication arrest p53 remains inactive in transcriptional transactivation, further supporting the direct involvement in HR regulatory functions unrelated to transcriptional transactivation activities.

p53 preferentially represses HR between certain mispaired DNA sequences. p53 specifically recognizes preformed heteroduplex joints structurally resembling early recombination intermediates, when comprising these mispairings [68]. p53 is able to attack DNA by 3′–5′ exonuclease activity principally during Rad51-mediated strand transfer and to display a DNA substrate preference for heteroduplex recombination intermediates with a further enhancement of the exonucleolytic activity for mispaired as compared to correctly paired heteroduplex DNA [38].

Highlighting the significance of p53 DNA interactions in the regulation of strand exchange events, p53 inhibits branch migration of Holliday junctions (HJs) [25, 31]. p53 recognizes this HJs -like structure and controls the generation and branch migration of the replication fork as well as its resolution, to prevent error-prone DSB repair and to cause replication pausing until the DNA lesion is repaired.

f. Mammalian cells repair the majority of double-strand breaks by NHEJ [69, 70] which is regarded as principally inaccurate process. The role of p53 in NHEJ remains unclear. p53 has an inhibitory effect on error-prone NHEJ but not error-free NHEJ [71], thereby suppressing genomic instability arising from low-fidelity repair. Remarkably, after the exposure to IR, DSB rejoining increases with loss of wtp53function. Inhibition of in vitro end-joining was observed with the oncogenic mutant p53(175H), whereas the phosphorylation-mimicking mutant p53(15D) failed to inhibit, thereby providing evidence for possible role of phosphorylated p53 in the regulation of NHEJ [72].

Various *in vitro* and *in vivo* studies have shown that p53 can rejoin or ligate compatible ends of DNA with DSBs [68, 70]. Evidently, p53 has several genetic interactions with components of the NHEJ pathway that are exhibited by downstream effects on cellular survival and cell-cycle control or effects on DNA repair. The molecular mechanisms of these interactions remain unresolved.

3.2 p53 and DNA repair in cytoplasm

Under normal conditions a basal pool of p53 is retained intra-cellular, with the distribution of p53 between the different subcellular compartments dependent on the cellular stress milieu [28]. Indeed, wtp53 occurs in cytoplasm in a subset of human tumor cells such as breast cancers, colon cancers and neuroblastoma [73–75]. Shuttling between nucleus and cytoplasm not only regulates protein localization, but also often impacts on protein function.

p53, localized in the cytoplasmic lysates of non-stressed p53-proficient cell lines [e.g. LCC2, HCT116 (p53+/+)] exerts an inherent 3′ → 5′ exonuclease activity displaying identical biochemical functions characteristic for recombinant wtp53 [76, 77]: 1) it removes 3′-terminal nucleotides from various nucleic acid substrates: ssDNA, dsDNA, and RNA/DNA template-primers, 2) it hydrolyzes ssDNA in preference to dsDNA substrate, 3) it shows a marked preference for excision of a mismatched vs. correctly paired 3′ terminus with RNA/DNA and DNA/DNA substrates, 4) it excises nucleotides from nucleic acid substrates independently from DNA polymerase, 6) it fulfills the requirements for proofreading function; acts coordinately with the exonuclease-deficient viral DNA polymerases.

Viruses exploits their cellular host for their successful replication, they utilize cell proteins for multiple purposes during their intracellular replication [78]. Since viral infection evokes cellular stress, the infected cells harbor stabilized activated p53 and manipulate p53's guardian role. Interestingly, increased p53 levels have been noted following infection of cells with various viruses including retrovirus-human immunodeficiency virus [79], which exhibits exceptionally high genetic variability [6], due to the low fidelity of the replication apparatus that is exclusive to the retroviral life cycle.

Reverse transcriptase (RT) of HIV-1 is responsible for the conversion of the viral genomic ssRNA into the proviral DNA in the cytoplasm [7]. The lack of intrinsic $3' \rightarrow 5'$ exonuclease activity, the formation of 3'-mispaired DNA and the subsequent extension of this DNA were shown to be determinants for the low fidelity of HIV-1 RT [80]. p53 can proofread for HIV-1 RT, increasing the fidelity of DNA synthesis by excising incorrectly polymerized nucleotides from RNA/DNA and DNA/DNA temple-primers in the direct exonuclease assay, when first binding to a 3'-terminus and during ongoing DNA synthesis *in vitro* with both template-primers [76]. The role of p53 in proofreading is two-fold: to excise preexisting 3'-terminal mismatches and to prevent the extension of 3'-mismatched primer ends by the polymerase [76]. p53 with its inherent exoribonuclease activity and excision of mispairs, has a potential to serve as an external trans-acting proofreader, providing the host-derived repair mechanism in cytoplasm.

3.3 p53 and DNA repair in mitochondria

DNA polymerase (pol) γ is the sole DNA polymerase that is responsible for replication and repair of mtDNA [81]. It is well established that defects in mtDNA replication lead to mitochondrial dysfunction and disease [56, 60]. Mutations in mtDNA can arise from exogenous sources, from endogenous oxidative stress, or as spontaneous errors of replication during either DNA synthesis or repair events [82]. Mitochondrial DNA is replicated by DNA polymerase γ in concert with replisome accessory proteins such as the mitochondrial DNA helicase, single-stranded DNA binding protein, topoisomerase, the multifunctional mitochondrial transcription factor A (TFAM) with important roles in mtDNA replication and initiating factors.

A high frequency of mutations within mtDNA, resulting in mitochondrial dysfunctions, is an important source of various diseases including cancer and human aging [81, 82]. To verify mtDNA integrity, cells hold various DNA damage response pathway(s) comprising mtDNA replication/repair preservation programs that either preclude or repair damage [83]. The mutagenic mechanisms were shown to be replication errors formed by either pol γ during DNA synthesis by incorporation of incorrect nucleotide or produced due to the presence of unbalanced dNTP concentrations, or by diminished proofreading efficiency. MtDNA is not protected by histones and mtDNA repair is ineffective [81]. Furthermore, a potentially important source of replication infidelity is damage due to ROS. pol γ, was demonstrated to stably misincorporate highly mutagenic 8-oxo-7,8-dihydro-2'-deoxyguanosine (8-oxodG) opposite template adenine in a complete DNA synthesis reaction *in vitro* [84].

Because of the susceptibility of mtDNA to oxidative damage and replication errors, it is vital to protect mtDNA genomic stability to preserve health. Mitochondrial localization of p53 was observed in non-stressed and stressed cells [26]. Mitochondrial p53 (mit-p53) levels are proportional to total p53 levels, and the majority of p53 was present inside the intra-mitochondrial compartment-matrix, in which mtDNA is located [85]. The mit-p53 physically and functionally interacts with both, mtDNA and pol γ [86].

Notably, with the exception of NER, components of these nuclear DNA repair pathways are also shared in mtDNA maintenance. Several studies illustrated the participation of p53 in mtDNA repair:

a. 53 enhances mitochondrial BER (mtBER) through direct interaction with the repair complex in mouse liver and cancer cells [87]. p53 modulates mtBER through the stimulation of the nucleotide incorporation step.

b. p53 interacts physically with human mtSSB (HmtSSB) *in vitro* via its transactivation domain and is proficient of hydrolyzing the 8-oxodG present at the 3′-end of DNA, a well-known marker of oxidative stress [88].

c. Intra-mitochondrial p53 provides an error-repair proofreading function for pol γ by excision of misincorporated nucleotides [89]. The p53 in mitochondria may affect the accuracy of DNA synthesis by acting as an external proofreader, thus reducing the production of polymerization errors.

4. Removal of nucleoside analogs from DNA by p53 protein

In addition to having a critical role in preservation of genome integrity, alterations in the expression, and function of DNA repair proteins are a major facilitator of tumor responses to chemo- and radiotherapy, commonly functioning by inducing DNA damage in tumor cells. Nucleoside analogs, clinically active in cancer chemotherapy (*e.g.* Ara-C, in the treatment of hematological malignancies, or gemcitabine-dFdC, against a variety of solid tumors) and in treatment of virus infections (*e.g.* 3′-azido-2,3,-deoxythymidine-AZT, inhibitors of HIV-1 RT), are incorporated into DNA and cause cell death or inhibition of viral replication [90, 91]. These drugs are intracellularly converted to the active analog triphosphates, compete with physiological nucleosides and are then inserted into replicating DNA. The incorporated NA, structurally mimicking a mismatched nucleotide at the 3′-terminus, blocks further extension of the nascent strand (chain termination) and causes stalling of replication forks with higher probability to the dissociation of the enzyme from template-primer [91]. Furthermore, the high toxicity of NA compounds may be caused by high rates of incorporation of the NA into DNA and their persistence in DNA due to inefficient excision. Removal of drugs by 3′ → 5′ exonuclease activity intrinsic to DNA polymerase or by external proofreading activity of external polymerases or proteins is presumably a potential cellular mechanism of resistance to anti-viral drugs or anti-cancer drugs.

The cytotoxic activity of gemcitabine (2′2′-difluorodeoxycitidine, dFdC) was strongly correlated with the amount of dFdCMP incorporated into cellular DNA [92]. The p53 protein recognizes dFdCMP-DNA in whole cells, as evidenced by the fact that p53 protein rapidly accumulated in the nuclei of the gemcitabine treated ML-1 cells [93]. Although, the excision of the dFdCMP from the 3′-end of the DNA was slower than the excision of mismatched nucleotides in whole cells with wtp53 (ML-1) and not detectable in CEM cells harboring mutant p53. ML-1 cells were more sensitive to the cytotoxic effect of the drugs compared to the p53-null or mutant cells. The recognition of the incorporated NAs in DNA by wtp53 did not confer resistance to gemcitabine, but may have facilitated the apoptotic cell death process. It was reported that treatment with gemcitabine resulted in an increased production of DNA-dependent protein kinase (DNA-PK) and p53 complex in nucleus, that interacts with the gemcitabine-containing DNA [93, 94]. DNA-PK and p53 sensor complex may serve as a mechanism to activate the pro-apoptosis function of p53. Apparently, the prolonged existence of the NA-stalled DNA end induced the kinase activity, which subsequently phosphorylated p53 and activated the downstream pathways leading to apoptosis.

Remarkably, p53 present in complex with DNA-PK exhibited 3′ → 5′ exonuclease activity with mismatched DNA, however the active p53 was unable of excising efficiently the incorporated drug from NA-DNA construct containing gemcitabine at the 3′-end [94]. Notably, the specific effects of gemcitabine exposure appeared to vary depending on the duration of treatment and upon the cell line.

It should be pointed out, that wtp53 in ML-1 cells removed the purine nucleoside analog fludarabine (F-ara-A) more efficiently than gemcitabine [93]. Further studies are needed to assess the role of p53 in cellular response to various anti-cancer purine and pyrimidine NA-induced DNA damage.

HIV-1 RT readily utilizes many NAs and the incorporation of nucleoside RT inhibitors (NRTIs) into the 3′-end of viral DNA leads to chain termination of viral DNA synthesis in cytoplasm [88, 95]. p53 protein in the cytoplasm excises the incorporated NAs during both RNA-dependent and DNA-dependent DNA polymerization reactions, although less efficiently than the mismatched nucleotides; longer incubation times were required for excision of the terminally incorporated analogs [96]. The data suggest that p53 in cytoplasm may act as an external proofreader for NA incorporation and confer cellular resistance mechanism to the anti-viral compounds.

Pol γ is unique among the cellular replicative DNA polymerases as it is sensitive to inhibition by nucleoside analogue reverse transcriptase inhibitors (NRTIs) used in the treatment of HIV, which can cause an induced mitochondrial toxicity [97]. Acquired mitochondrial toxicity occurs as a consequence of incorporation of NA into mtDNA or inhibition of mtDNA replication or both. A terminally incorporated NA may be removed by p53 in mitochondria [97]. The removal of the incorporated NA by p53 exonuclease, indicates that the presence of the cellular component-p53 in mitochondria may be important in defining the cytotoxicity of NAs toward mitochondrial replication, thus affecting risk–benefit approach (NA toxicity versus viral inhibition) [98, 99]. Apparently, the presence of p53 in mitochondria may be important, as the excision of the mispair and NA by p53 is favorable event for mitochondrial function.

p53 is a multifunctional protein with positive and negative effects. In general, drug resistance that occurs in cancer chemotherapy and antiviral therapy is a negative event that will decrease the efficacy of the treatment. The recognition and removal of NA from drug-containing DNAs by p53 exonuclease activity in various compartments of the cell may play a role in decreasing drug activity, leading to various biological outcomes: 1)the excision of the incorporated NA from DNA in nucleus may confer resistance to the drugs (negative effect) [93]; 2)the removal of the NA by p53 from DNA incorporated by HIV-1 RT in cytoplasm may confer resistance to the drugs by non-viral mechanism (negative effect) [96] and 3)the excision of NAs from mitochondrial DNA may decrease the potential for chain termination and host toxicity (positive effect) [97].

5. Excision of non-canonical nucleotides by p53 protein

The genome is constantly under attack from extrinsic and intrinsic damaging agents. Uracil (dU) mis-incorporation in DNA is an intrinsic factor resulting in genomic instability and DNA mutations. The excessive levels of genomic uracil in DNA can modify gene expression by interfering with promoter binding and transcription inhibition, can change transcriptional stalling, or induce DNA strand breaks leading to apoptosis. The factors that influence uracil levels in DNA are cytosine deamination, de novo thymidylate (dTMP) biosynthesis, salvage dTMP biosynthesis, and DNA repair. Furthermore, mis-incorporation occurs when DNA polymerases incorporate dUTP into DNA, in place of dTTP, and the rate of mis-incorporation is believed to be determined by the intracellular dUTP:dTTP ratio [100, 101]. The enzyme deoxyuridine triphosphate nucleotidohydrolase (dUTPase), which facilitates the conversion of dUTP to dUMP further utilized by thymidylate synthase (TS) for synthesis of dTMP, avoids mis-incorporation of dU into DNA in nucleus by decreasing the dUTP/dTTP ratio [101]. The misincorporation of dU, as a result of accumulation of dUTP, plays a critical role in cytotoxicity mediated by TS inhibitors, such as the commonly used anticancer drug 5-fluorouracil (5-FU) [102]. DNA directed cytotoxicity of chemotherapeutic agents (e.g.5-FU) not only depends on accumulation of dUTP, but may also be determined by the efficiency of the DNA repair mechanisms (e.g. excision repair) which preclude the incidence of the mistake.

Pol γ in mitochondria is incapable to readily correct U:A mismatches [11]. HIV-1 RT in the cytoplasm of HIV-infected cells efficiently inserts the non-canonical dUTP into the proviral DNA and extends the dU-terminated DNA [103]. The misincorporation of dUTP leads to mutagenesis, and to down-regulation of viral gene expression [104].

Within the context of error-correction events, p53 as a DNA binding protein, contributes an external proofreading function; upon excision of the dU, the p53 dissociates, thus letting the transfer of the substrate with the correct 3′-terminus to DNA polymerase and renewal of DNA synthesis.

The biochemical data show that the procession of U:A and mismatched U:G lesions enhances in the presence of recombinant or endogenous cytoplasmic or mitochondrial p53 [105]. p53 in cytoplasm can participate through the intermolecular pathway in a dU-damage-associated repair mechanism by its ability to remove preformed 3′-terminal dUs, thus preventing further extension of 3′ dU-terminated primer during DNA synthesis by HIV-1 RT. Similarly, p53 in mitochondria can function as an exonuclease/proofreader for pol γ by either decreasing the incorporation of non-canonical dUTP into DNA or by promoting the excision of incorporated dU from nascent DNA, thus expanding the spectrum of DNA damage sites exploited for proofreading as a trans-acting protein [106].

During genomic DNA replication another form of replication errors arises during the incorporation of nucleotides carrying the correct base, but the wrong sugar at substantial rates [107]. DNA polymerases often incorporate ribonucleoside triphosphates (rNTPs) into DNA because of the much higher concentration of rNTPs than that of dNTPs in the cellular nucleotide pool. Indeed, more than 10^6 rNMPs are incorporated during one round of replication of a mammalian genome [107]. Newly incorporated rNMPs destabilize DNA and pose a major threat to genome integrity due to their reactive 2'OH group. The inserted rNs are the most abundant non-canonical nucleotides in the genome. Failure of rN removal is associated with genome instability in the form of mutagenesis, replication stress, DNA breaks, and chromosomal rearrangements. The aberrant accumulation of rNs in the genome leads to human diseases including Aicardi–Goutières syndrome (AGS), the severe autoimmune disease, and tumorigenesis [108]. Mammalian cells have developed strategies to prevent persistent rN accumulation. In eukaryotes, rNs embedded into DNA are primarily repaired by RNase H2-initiated repair pathway. Ribonucleotide excision repair (RER) may be directly coupled to replication and results in rapid post-replicative repair of rNMPs [108]. Remarkably, exonuclease-proficient yeast and human DNA polymerases can proofread incorporated rNs, albeit inefficiently [107].

Recent studies have demonstrated the importance of p53 in 3′-terminal RER pathway through a functional collaboration with HIV-1 RT, acting in a coordinated manner to attain higher fidelity. p53, functioning as a trans-acting proofreader in cytoplasm, can decrease the stable incorporation of rNs, into DNA by HIV-1 RT [109]. p53 can influence events needed for RER by possessing the compatible biochemical properties: p53 is pertinent in the correction of replication errors produced by HIV-1 RT during distinct steps of rN incorporation through intermolecular pathway: by removal pre-existing 3′-terminal rN; by reducing rN incorporation; by preventing extension of a 3′ rN-terminated primer, by attenuating stable incorporation of rNs. Thus, p53, functioning as a trans-acting proofreader in cytoplasm, can decrease the stable incorporation of rNs.

The fact that p53 in cytoplasm can edit an incorrect sugar irrespective of the nature of base, expands the role of p53 as a proofreader in the repair of replication errors by removing both a base mismatch and an incorrect sugar.

6. Conclusions

Mammalian cells have evolved multiple strategies to safeguard the genetic information to prevent the fixation of genetic damage induced by endogenous and exogenous mutagens [16]. p53 protein plays a crucial role in the regulation of cell fate determination in response to a variety of cellular stresses. p53 may exert the functional heterogeneity in its non-induced and in its activated state [16]. Remarkably, DNA repair transcription-independent functions of wtp53, contributing to tumor suppression, were found to protect cells from DNA damage independently of the transcription-mediated functions of p53 [25]. Thus, a more comprehensive understanding of how p53 transcription- independent functions are induced in response to a variety of cellular insults is vital. This report focuses on direct roles of p53 in DNA repair during DNA replication in various compartments of the cell. Apparently, p53 has more than one contributions to DNA replication fidelity, which could depend on sub-cellular localization of p53, on the type and incidence of replication obstacles, on the levels of p53 protein [28].

p53 is able to elicit a spectrum of different effective DNA repair pathways in nucleus, cytoplasm and mitochondria (**Figure 4**). Within the nucleus, p53 regulates different repair mechanisms, in response to endogenous and exogenous replicative stress: *e.g.* HR (by restricting excess recombination through interactions with Rad51), NER, BER, and MMR through interactions with relevant components of the respective pathways [25, 31].

In the cytoplasm, p53 may contribute effective proofreading for exonuclease-deficient DNA polymerases (*e.g.* HIV-1 RT) thereby correcting errors produced during DNA replication [110, 111]. Moreover, the proofreading activity of p53 may limit the transversion mutations, indicating that p53 may affect the mutation spectra of DNA polymerase by acting as an external proofreader [111]. Recent studies also show that cytoplasmic p53 possesses the potential to remove the incorporated non-canonical dUTP into DNA by HIV-1 RT through an intermolecular pathway [105]. Furthermore, p53, functioning as a trans-acting proofreader, can decrease the stable incorporation of rNs [109]. The data implies that p53 excises incorrect sugar in addition to base mispairs, thereby expanding the role of p53 in the repair of replication errors.

Within the mitochondria, various studies illustrated the participation of p53 in mtDNA repair in a variety of systems: a) p53 enhances BER through direct interaction with the repair complex in mouse liver and cancer cells [87]. b) Intra-mitochondrial p53 provides an error-repair proofreading function for pol γ by

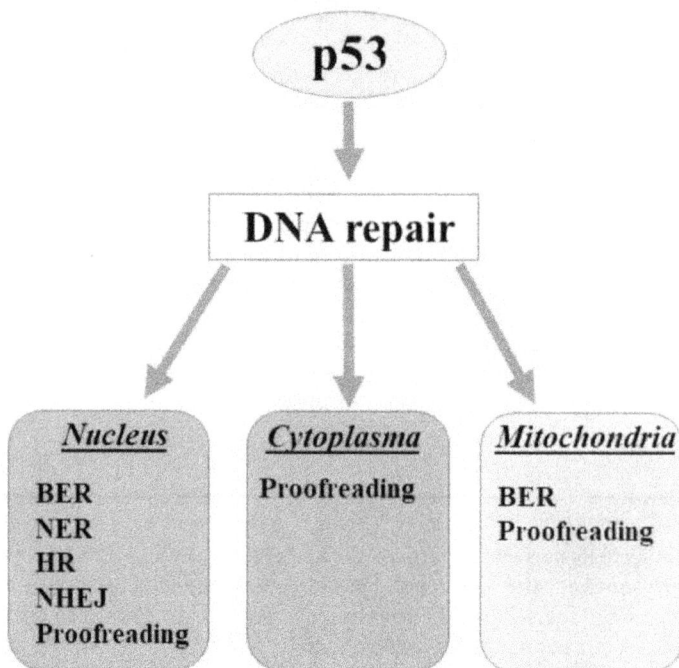

Figure 4.
p53 functions in DNA repair. p53 under both normal and stress conditions, can help cellular and viral DNA polymerases to promote the repair of DNA in various cellular compartments. The result of p53 activation depends on many variables, including the extent of the stress or damage. In this model, basal p53 activity or that induced by stress signals elicits the protector responses that support the repair of genotoxic damage by various pathways.

excision of misincorporated nucleotides [89]. c)p53 is proficient of hydrolyzing the 8-oxo-7,8-dihydro-2'-deoxy-guanosine (8-oxodG) present at the 3'-end of DNA, a well-known marker of oxidative stress [88]. d)p53 regulates mtDNA copy number, which may impact mitochondrial and cellular functions [112].

Therapeutic strategies based on p53 are particularly interesting because they exploit the cancer cell's intrinsic genome instability and predisposition to cell death-apoptosis [90, 91]. The role of p53 is predominantly relevant with respect to the development of anticancer and antiviral therapies. Removal of drugs by 3' → 5' exonuclease activity may also facilitate resistance to anti-cancer or anti-viral treatments. Clinical drug resistance limits the efficacy of these compounds. Uncovering the mechanisms, which are responsible for DNA repair of NA-induced DNA damage will have therapeutic value. The p53 protein is able to remove incorporated NA. The stress induced activation of p53 that occurs during anti-cancer or anti-viral therapy has negative and positive effects. p53 may remove incorporated therapeutic NAs from DNA or trigger apoptosis. More studies regarding functions of p53 in genome integrity and cancer evolution may facilitate drug screening and better design of therapeutic approaches.

7. Future directions

The functional interaction between p53 and DNA polymerase may have important consequences for the maintenance of genomic integrity and in the development of p53- targeted clinical therapies. Further assessments are required to establish the role of p53 in DNA replication and the significance of these functions in various cellular compartments and treatment responses. Studies on the biology of various mutant p53 isoforms and their interaction with the factors involved in DNA repair and apoptosis, will be relevant to establish whether the direct involvement of p53 in DNA repair is a tumor suppressor function of this important anti-oncogene. Characterization of exonuclease-deficient H115N mutant p53 revealed that although exonuclease-mutant H115N p53 can induce cell cycle arrest more efficiently than wild-type p53, its ability to produce apoptosis in DNA damaged cells is markedly impaired [113]. By utilizing various function-mutant p53 isoforms, more studies must be conducted on the biology of mutant p53 forms and their interaction with the factors involved in DNA repair and apoptosis, in order to recognize the molecular mechanisms that mediate p53-dependent control of DNA replication by cellular and viral DNA polymerases.

p53 has a dual role in response to therapy, as exonuclease that by excision of incorporated anti-cancer drugs may confer resistance to drugs or as mediator of cell death induced by chemotherapy [93]. p53, by removal of the incorporated NA, could confer a cellular resistance mechanism to the antiviral compounds. Finally, the excision of NAs from mitochondrial DNA may decrease the potential for chain termination and host toxicity. These features could serve as a template for the development of p53-targeting therapies.

The control of the viral mutation rate could be a practical anti-retroviral strategy. The mutagenic capacity of a low fidelity DNA polymerase will be decreased through increase in exonuclease concentration or exonuclease targeting (increase in local p53 concentration). It is important to further elucidate the molecular mechanisms involved in governing fidelity not only at a molecular level (*i.e.*, intrinsic RT fidelity), but also related to the cytoplasmic p53 protein that can control the viral mutation rate and can affect the incorporation of NAs into viral DNA. New understandings of the sub-cellular localization of p53, its role in the fidelity of proviral DNA synthesis in cytoplasm and drug resistance, may create the basis for new strategies in targeted antiviral therapy that focus on the sub-cellular context of p53 in cells.

A major issue in the future would be to characterize the cellular and biological functions of p53 in mitochondria in response to various stresses. There are many missing links about the biological functions of mitochondrial p53 that are required to be investigated. Whether p53 defines the percent of mutated mtDNA (heteroplasmy in a cell)? Uncovering the mechanisms by which pol γ-mediated mtDNA mutations and depletion are manifested in cells in the absence and presence of p53 is significant step in understanding underlying causes for mtDNA–related diseases. Depletion and mutation of mtDNA may lead to cellular respiratory dysfunction and release of reactive oxidative species, resulting in cellular damage [99]. Future NAs should provide higher specificity for HIV-RT and lower incorporation by pol γ to diminish mitochondrial toxicity. Whether the effective targeting of p53 in mitochondria by error-correction functions, may result in decrease of mitochondrial toxicity in response to conventional anti-viral therapies? Understanding how p53 can be imported into mitochondria, will be important and could contribute toward the design of new therapies for various diseases.

Author details

Bakhanashvili Mary[1,2]

1 Infectious Diseases Unit, Sheba Medical Center, Tel-Hashomer, Israel

2 The Mina and Everard Goodman Faculty of Life Sciences, Bar-Ilan University, Ramat-Gan, Israel

*Address all correspondence to: bakhanus@yahoo.com

IntechOpen

References

[1] Kunkel TA, Bebenek K. (2000) DNA replication fidelity. Ann. Rev. Biochem. 69: 497-529.

[2] Lan-Ya Li, Yi-di Guan , Xi-Sha Chen, Jin-Ming Yang , Yan Cheng. (2021) DNA repair pathways in cancer therapy and resistance. Front Pharmacol. 11: 629266.

[3] D'Amico AM, Vasquez KM. (2021) The multifaceted roles of DNA repair and replication proteins in aging and obesity. DNA Repair 99:103049.

[4] Rahimian E, Amini A, Alikarami F, Pezeshki SMS, Saki N, Safa M. (2020) DNA repair pathways as guardians of the genome: Therapeutic potential and possible role in hematologic neoplasms. DNA repair 96:102951.

[5] Jackson AL, Loeb LA. (1998) The mutation rate and cancer. Genetics 148: 1483-1490.

[6] Katz R, Skalka AM. (1990) Generation of diversity in retroviruses. Ann. Rev. Genet. 24: 409-445.

[7] Svarovskaya ES, Cheslock SR, Zhang W, Hu W, Pathak VK (2003) Retroviral mutation rates and reverse transcriptase fidelity. Front. Biosci.8: d117-d134.

[8] Menéndez-Arias L. (2009) Mutation rates and intrinsic fidelity of retroviral reverse transcriptases. Viruses. 1: 1137-1165.

[9] Trimpert J, Groenke N, Kunec D, Eschke K, He S, McMahon DP, Osterrieder N. (2019) A proofreading-impaired herpersvirus generates populations with quasispecies-like structure. Nat. Microbiology 4:2175-2183.

[10] Copeland WC, Ponamarev MV, Nguyen D, Kunkel TA, Longley MJ. (2003) Mutations in DNA polymerase gamma cause error-prone DNA synthesis in human disorders. Acta Biochim Pol. 50:155-167.

[11] Kaguni LS. (2004) DNA polymerase γ, the mitochondrial replicase. Ann. Rev. Biochem.73:293-320

[12] Singh KK. (2004) Mitochondria damage checkpoint in apoptosis and genome stability. FEMS Yeast Res.5: 127-132.

[13] Pesole G, Gissi C, De Chirico A, Saccone C. (1999) Nucleotide substitution rate of mammalian mitochondrial genomes. J Mol Evol 48: 427-434.

[14] Johnson A, Johnson K. (2001a) Fidelity of nucleotide incorporation by human mitochondrial DNA polymerase. J Biol. Chem. 276: 38090-38106.

[15] Johnson A, Johnson K. (2001b) Exonuclease proofreading by human mitochondrial DNA polymerase. J Biol. Chem. 276: 38097-38107.

[16] Zhang H. (2020) Mechanisms of mutagenesis induced by DNA lesions: multiple factors affect mutations in translesion DNA synthesis. Crit Rev Biochem. Mol Biol. 55:219-251.

[17] Reha-Krantz. (2010) DNA polymerase proofreading: Multiple roles maintain genome stability. BBA 1804: 1049-1063.

[18] McElhinny SAN, Pavlov Y, Kunkel T. (2006) Evidence for extrinsic exonucleolytic proofreading. Cell Cycle 5: 958-962.

[19] Levine AJ, Oren M. (2009) The first 30 years of p53: growing ever more complex. Nat Rev Cancer. 9:749-758.

[20] Vousden KH, Prives C (2009) Blinded by the light: The growing complexity of p53. Cell 137: 413-431.

[21] Soussi T (1995) The p53 tumor suppressor gene: From molecular biology to clinical investigation. In Molecular Genetics of Cancer (Cowell, J.K., ed.), p135-178, Bios. Scientific, Oxford, UK.

[22] Freed-Pastor WA, Prives C. (2012) Mutant p53: One name, many proteins. Genes. Dev. 26: 1268-1286.

[23] Levine AJ. Levine AJ. (2020) p53:800 million years of evolution and 40 years of discovery. Nat Rev Cancer.20: 471-480

[24] Taira N, Yoshoda K. (2012) Post-translational modifications of p53 tumor suppressor:Determinants of its functional targets. Histol Histopathol. 27: 437-443.

[25] Ho T, Tan BX, Lane D. (2019) How the other half lives: What p53 does when it is not being a transcription Factor. Int J Mol Sci. 21: 13.

[26] Marchenko ND, Zaika A, Moll UM. (2000) Death signal-induced localization of p53 protein to mitochondria. A potential role in apoptotic signaling. J. Biol. Chem. 275: 16202-16212.

[27] Zhou J, Ahn J, Wilson SH, Prives C. (2001) A role for p53 in base excision repair. EMBO J. 20: 914-923.

[28] Janus F, Albrechtsen N, Knippschild U, Wiesmuller L, Grosse F, Deppert W. (1999) Different regulation of the p53 core domain activities 3′ to 5′ exonuclease and sequence-specific DNA binding. Mol. Cell. Biol. 19: 2155-2168.

[29] Offer H, Wolkowicz R, Matas D, Blumenstein S, Livneh Z, Rotter V. (1999) Direct involvement of p53 in the base excision repair pathway of the DNA repair machinery. FEBS Lett. 450: 197-204.

[30] Albrechtsen N, Dornreiter L, Grosse F, Kim E, Wiesmuller L, Deppert W (1999) Maintenance of genomic integrity by p53: Complementary roles for activated and non-activated p53. Oncogene 18: 7706-7717.

[31] Gatz SA, Wiesmüller L. (2006) p53 in recombination and repair. Cell Death Differ;13:1003-1016.

[32] Gottifredi V, Wiesmüller L. (2018) The tip of an iceberg: Replication-associated functions of the tumor suppressor p53. Cancers 10: 250

[33] Kern SE, Kinzler KW, Baker SJ, Nigro JM, Rotter V, Levine AJ, Friedman P, Prives C. Vogelstein B. (1991) Mutant p53 binds DNA abnormally. Oncogene 6: 131-136.

[34] Steinmeyer K, Deppert W. (1988) DNA binding properties of murine p53. Oncogene 3: 501-507.

[35] Bakalkin G, Yakovleva T, Selivanova G, Magnusson KP, Szekely L, Kiseleva E, Klein G, Terenius L, Wiman KG. (1994) p53 binds single-stranded DNA ends and catalyzes DNA renaturation and strand transfer. Proc. Natl. Acad. Sci. USA 91: 413-417.

[36] Oberosler P, Hloch P, Rammsperger U, Stahl H. (1993) p53-catalyzed annealing of complementary single-stranded nucleic acids. EMBO J 12: 2389-2396.

[37] Lee S, Elenbaas B, Levine A, Griffith J (1995) p53 and its 14kDa C-terminal domain recognize primary DNA damage in the form of insertion/deletion mismatches. Cell 81: 1013-1020.

[38] Dudenhoffer C, Rohaly G, Will K, Deppert W, Wiesmullar L.(1998) Specific mismatch recognition in heteroduplex intermediates by p53 suggests a role in fidelity control of homologous recombination. Mol.Cell. Biol. 18: 5332-5342.

[39] Hwang BJ, Ford J.M, Hanawalt PC, Chu G. (1999) Expression of the p48

xeroderma pigmentosum gene is p53-dependent and is involved in global genomic repair. Proc.Natl. Acad. Sci. USA, 96: 424-428.

[40] Mummenbrauer T, Janus F, Muller B, Wiesmuller L, Deppert W, Gross F. (1996) p53 protein exhibits 3'→5' exonuclease activity. Cell 85: 1089-1099.

[41] Huang P. (1998) Excision of mismatched nucleotides from DNA: A potential mechanism for enhancing DNA replication fidelity by the wild-type p53 protein. Oncogene 17: 261-270.

[42] Skalski V, Lin Z, Choi BY, Brown KR. (2000) Substrate specificity of the p53-associated 3'→5' exonuclease. Oncogene 19: 3321-3329.

[43] Bakhanashvili M. (2001) Exonucleolytic proofreading by p53 protein. Eur.J Biochem. 268: 2047-2054.

[44] Offer H, Milyavsky M, Erez N, Matus D, Zurer I, Harris CC, Rotter V. (2001) Structural and functional involvement of p53 in BER in vitro and in vivo. Oncogene 20: 581-589.

[45] Gaiddon C, Moorthy NC, Prives C. (1999) Ref-1 regulates the transactivation and proapoptotic functions of p53 in vivo. EMBO J. 18: 5609-5621.

[46] Williams, AB, Schumacher B. (2016) p53 in the DNA-damage-repair process. Cold Spring Harb. Perspect. Med. 6, a026070.

[47] Rubbi C.P, Milner J. (2003) p53 is a chromatin accessibility factor for nucleotide excision repair of DNA damage. EMBO J. 22, 975-986.

[48] Wang XW, Yeh H, Schaeffer L, Roy R, Moncollin V, Egly JM, Wang Z, Taffe BG, et al. (1995) p53 modulation of TFIIH-associated nucleotide excision repair activity. Nat. Genet. 10: 188-195.

[49] Scherer SJ, Maier SM, Seifert M, Hanselmann RG, Zang KD, Muller-Hermelink HK, Angel P, Welter C, Schartl M. (2000) p53 and c-Jun functionally synergize in the regulation of the DNA repair gene hMSH2 in response to UV. J. Biol. Chem. 275, 37469-37473.

[50] Subramanian D, Griffth JD.(2002) Interactions between p53, hMSH2-hMSH6 and HMG I(Y) on Hollidayjunctions and bulged bases. Nucleic Acids Res. 30, 2427-2434.

[51] Subramanian D, Griffth JD. (2005) Modulation of p53 binding to Holliday junctions and 3'-cytosine bulges by phosphorylation events. Biochemistry 44, 2536-2544.

[52] Cranston A, Bocker T, Reitmair A, Palazzo J, Wilson T, Mak T, Fishel R. (1997) Female embryonic lethality in mice nullizygous for both Msh2 and p53. Nat Genet. 17:114-118.

[53] Pabla N, Ma Z, McIlhatton MA, Fishel R, Dong Z. (2011) hMSH2 recruits ATR to DNA damage sites for activation during DNA damage-induced apoptosis. J Biol Chem. 286:10411-10418.

[54] Shevelev IV, Hubscher U. (2002) The 3'→5' exonucleases. Nat.Rev. 3: 1-12.

[55] Perrino FW, Preston BD, Sandell LL, Loeb LA. (1989) Extension of mismatched 3'termini of DNA is a major determinant of the infidelity of human immunodeficiency virus type 1 reverse transcriptase. Proc. Natl. Acad. Sci. USA 86: 8343-8347.

[56] Kunkel TA. (2004) DNA replication fidelity. J Biol. Chem. 279: 16895-16898.

[57] Bebenek K, Roberts JD, Kunkel TA. (1992) The effects of dNTP pool imbalances on frameshift fidelity during DNA replication. J Biol Chem.;267: 3589-3596.

[58] Boyer JC, Thomas DC, Maher VM, McCormick JJ, Kunkel TA. (1993) Fidelity of DNA replication by extracts of normal and malignantly transformed human cells. Cancer Res. 53: 3270-3275.

[59] Hurwitz J, Dean FB, Kwong AD, Lee SH. (1990) The in vitro replication of DNA containing the SV40 origin. J Biol Chem. 265:18043-18046.

[60] McCulloch SD, Kunkel TA. (2008) The fidelity of DNA synthesis by eukaryotic replicative and translesion synthesis polymerases. Cell Res. 18: 148-161.

[61] Cox LS, Hupp T, Midgley CA, Lane DP (1995) A direct effect of activated human p53 on DNA replication. EMBO J 14: 2099-2105.

[62] Ballal K, Zhang W, Mukhopadyay T, Huang P. (2002) Suppression of mismatched mutation by p53: A mechanism guarding genomic integrity. J. Mol. Med. 80: 25-32.

[63] Melle C, Nasheuer H. (2002) Physical and functional interactions of the tumor suppressor protein p53 and DNA polymerase α-primase. Nucleic Acids Res. 30: 1493-1499.

[64] Bertrand P, Saintigny Y, Lopez BS. (2004) p53Os double life: Transactivation-independent repression of homologous recombination. Trends Genet. 20, 235-243.

[65] Romanova LY, Willers H, Blagosklonny MV, Powell SN. (2004) The interaction of p53 with replication protein a mediates suppression of homologous recombination. Oncogene 23: 9025-9033.

[66] Rieckmann T, Kriegs M, Nitsch L, Hoffer, K, Rohaly G, Kocher S, Petersen C, Dikomey E, Dornreiter I, Dahm-Daphi J. (2013) p53 modulates homologous recombination at I-SceI-induced double-strand breaks through cell-cycle regulation. Oncogene 32: 968-975

[67] Janz C, Wiesmuller L. (2002) Wild-type p53 inhibits replication-associated homologous recombination. Oncogene 21: 5929-5933

[68] Linke SP, Sengupta S, Khabie N, Jeffries BA, Buchhop S, Miska S, et al. (2003) p53 interacts with hRAD51 and hRAD54, and directly modulates homologous recombination. Cancer Res. 63: 2596-2605.

[69] Yang T, Namba H, Hara T, Takmura N, Nagayama Y, Fukata S, Ishikawa N, Kuma K, Ito,.K, Yamashita S. (1997) p53 induced by ionizing radiation mediates DNA end-jointing activity, but not apoptosis of thyroid cells. Oncogene 14: 1511-1519.

[70] Tang W, Willers H, Powell SN. (1999) p53 directly enhances rejoining of DNA double-strand breaks with cohesive ends in gamma-irradiated mouse fibroblasts. Cancer Res. 59: 2562-2565.

[71] Akyuz N, Boehden GS, Susse S, Rimek A, Preuss U, Scheidtmann KH, Wiesmuller L. (2002) DNA substrate dependence of p53-mediated regulation of double-strand break repair. Mol. Cell. Biol. 22: 6306-6317.

[72] Okorokov AL, Warnock L, Milner J (2002) Effect of wild-type, S15D and R175H p53 proteins on DNA end-joining in vitro: Potential mechanism of DNA double-strand break repair modulation. Carcinogenesis 23: 549-557

[73] Stenmark-Askmalm M, Stal O, Sullivan S, Ferraud L, Sun XF, Carstensen J, Nordenskjold B. (1994) Cellular accumulation of p53 protein: An independent prognostic factor in stage II breast cancer. Eur. J Cancer 30A: 175-180.

[74] Moll UM, LaOuglia M, Benard, Riou G. (1995) Wild-type p53 protein undergoes cytoplasmic sequestration in undifferentiated neuroblastomas but not in differentiated tumors. Proc. Natl. Acad. Sci. US, 92: 4407-4411.

[75] Bosari S, Viale G, Roncalli M, Graziani D, Borsani G, Lee AK, Coggi G. (1995) p53 gene mutations, p53 protein accumulation and compartmentalization in colorectal adenocarcinoma. Am. J Pathol. 147: 790-798.

[76] Bakhanashvili M. (2001) p53 enhances the fidelity of DNA synthesis by human immunodeficiency virus type 1 reverse transcriptase. Oncogene 20: 7635-7644.

[77] Bakhanashvili M, Gedelovich R, Grinberg S, Rahav G. (2008) Exonucleolytic degradation of RNA by the tumor suppression protein p53 in cytoplasm. J Molec. Medicine 86: 75-88.

[78] Aloni-Grinstein R, Charni-Natan M, Solomon H, Rotter V. (2018) p53 and the viral connection: Back into the future. Cancers. 10 :178.

[79] Genini D, Sheeter D, Rought S, Zaunders JJ, Susin SA, Kroemer G, Richman DD, Carson DA, Corbeil J, Leoni LM. (2001) HIV induces lymphocyte apoptosis by a p53-initiated, mitochondrial- mediated mechanism. FASEB J. 15: 5-6.

[80] Bakhanashvili M, Hizi A (1992) Fidelity of the reverse transcriptase of human immunodeficiency virus type. FEBS Lett. 306: 151-156.

[81] Stumpf JD, Copeland WC. (2011) Mitochondrial DNA replication and disease: Insights from DNA polymerase γ mutations. Cell Mol Life Sci. 68: 219-233.

[82] Wallace DC. (2010) Mitochondrial DNA mutations in disease and aging. Env. Mol. Mutagenesis 51:440-450.

[83] Singh KK, Kulawiec M. (2009) Mitochondrial DNA polymorphism and risk of cancer. Methods Mol Biol. 471: 291-303.

[84] Katafuchi A, Nohmi T. (2010) DNA polymerases involved in the incorporation of oxidized nucleotides into DNA: Their efficiency and template base preference. Mut. Res. 703: 24-31.

[85] Mahyar-Roemer M, Fritzsche C, Wagner S, Laue M, Roemer K. (2004) Mitochondrial p53 levels parallel total p53 levels independent of stress response in human colorectal carcinoma and glioblastoma cells. Oncogene 23: 6226-6236.

[86] Achanta G, Sasaki R, Feng L, Carew JS, Lu W, Pelicano H, et al. (2005) Novel role of p53 in maintaining mitochondrial genetic stability through interaction with DNA pol γ. EMBO J 24: 3482-3492.

[87] de Souza-Pinto NC, Harris CC, Bohr VA. (2004) p53 functions in the incorporation step in DNA base excision repair in mouse liver mitochondria. Oncogene 23: 6559-6568.

[88] Wong TS, Rajagopalan S, Townsley FM, Freund SM, Petrovich M, Loakes D, Fersht AR. (2009) Physical and functional interactions between human mitochondrial single-stranded DNA binding protein and tumor suppressor p53. Nucleic Acids Res. 37:568-581.

[89] Bakhanashvili M. Grinberg S, Bonda E, Simon AJ, Moshitch-Moshkovitz S, Rahav G. (2008) p53 in mitochondria enhances the accuracy of DNA synthesis. Cell Death Diff. 15: 1865-1874.

[90] Keating MJ. (1997) In: Nucleoside Analogs in Cancer Therapy. Cheson BD. Keating, Plunkett W. (eds). Marcel Dekker, Inc., New York, pp201-226.

[91] Sluis-Cremer N, Arion D, Parniak MA. (2000) Molecular mechanisms of HIV-1 resistance to nucleoside reverse transcriptase inhibitors (NRTIs). Cell. Mol. Life Sci. 57:1408-1422.

[92] Zhou Y, Achanta G, Pelicano H, Gadhi V, Plunkett W, Huang P. (2002) Action of (E)-2'-Deoxy-2'- (fluoromethylene) cytidine on DNA metabolism: Incorporation, excision and cellular response. Mol. Pharmacology 61: 222-229.

[93] Feng L, Achanta G, Pelicano H, Zhang W, Plunkett W, Huang P. (2000) Role of p53 in cellular response to anticancer nucleoside analog-induced DNA damage. Int. J Molec. Medicine 5: 597-604.

[94] Achanta G, Pelicano H, Feng L, Plunkett W, Huang P. (2001) Interaction of p53 and DNA-PK in response to nucleoside analogues: Potential role as a sensor complex for DNA damage. Cancer Res 61: 8723-8729.

[95] Sluis-Cremer N, Arion D, Parniak MA. (2000) Molecular mechanisms of HIV-1 resistance to nucleoside reverse transcriptase inhibitors (NRTIs). Cell. Mol. Life Sci. 57:1408-1422.

[96] Bakhanashvili M, Novitsky E, Rubinstein E, Levy I, Rahav G. (2005) Excision of nucleoside analogs from DNA by p53 protein, a potential cellular mechanism of resistance to inhibitors of human immunodeficiency virus type 1 reverse transcriptase. Antimic. Agents and Chem. 49:1576-1579.

[97] Bakhanashvili M, Grinberg S, Bonda E, Rahav G. (2009) Excision of nucleoside analogs in mitochondria by p53 protein. AIDS 23: 779-788.

[98] Fowler JD, Brown JA, Johnson KA, Suo Z. (2008) Kinetic investigation of the inhibitory effect of gemcitabine on DNA polymerization catalyzed by human mitochondrial DNA polymerase. J Biol. Chem. 283: 15339-15348.

[99] Lewis W, Day BJ, Copeland WC. (2003) Mitochondrial toxicity of NRTI antiviral drugs: An integrated cellular perspectives. Nature Reviews 2: 812-822.

[100] Olinski,R, Jurgowiak M, Zaremba TZ. (2010) Uracil in DNA–its biological significance. Mut. Res. 705 239-245.

[101] Verri A, Mazzarello O, Spadari S, Focher R. (1992) Uracil-DNA glycosylases preferentially excise mispaired uracil. Biochem. J 287: 1007-1010.

[102] Longley DB, Harkin DP, Johnston PG. (2003) 5-fluorouracil: Mechanisms of action and clinical strategies. Nat.Rev. Cancer. 3: 330-338.

[103] Kennedy EM, Daddacha W, Slater R, et al. (2011) Abundant non-canonical dUTP found in primary human macrophages drives its frequent incorporation by HIV-1 reverse transcriptase. J Biol. Chem. 286: 25047-25055.

[104] Hansen EC, Ransom M, Hesselberth JR. et al. (2016) Diverse fates of uracilated HIV-1 DNA during infection of myeloid lineage cells. Elife 5 e18477.

[105] Saragani Y, Hizi A, Rahav G, Zauch S, Bakhanashvili M. (2018) Cytoplasmic p53 Contributes to the Removal of Uracils Misincorporated by HIV-1 Reverse Transcriptase. Biochem Biophys Res Commun. 497: 804-810.

[106] Bonda E, Rahav G, Kaya A, Bakhanashvili M. (2016) p53 in the mitochondria, as a trans-acting protein, provides error-correction activities during the incorporation of non-canonical dUTP into DNA. Oncotarget 7: 73323-73336.

[107] Zhou ZX, Williams JS, Lujan SA, Kunkel TA. (2021) Ribonucleotides incorporation into DNA during DNA replication and its consequences. Crit Rev Biochem. Mol. Biol. 56:109-124.

[108] Cerritelli SM, El Hage A. (2020) RNases H1 and H2: guardians of the stability of the nuclear genome when supply of dNTPs is limiting for DNA synthesis. Curr Genet. 66:1073-1084.

[109] Akua T, Rahav G, Saragani Y, Hizi A, Bakhanashvili M. (2017) Tumor suppressor p53 protein removes ribonucleotides from DNA incorporated by HIV-1 reverse transcriptase. AIDS 31:343-353.

[110] Lilling G, Novitsky E, Sidi Y, Bakhanashvili M. (2003) p53-associated $3' \rightarrow 5'$ exonuclease activity in nuclear and cytoplasmic compartments of the cells. Oncogene 22: 233-245.

[111] Bakhanashvili M, Novitsky E, Lilling G, Rahav G. (2004) p53 in cytoplasm may enhance the accuracy of DNA synthesis by human immunodeficiency virus type 1 reverse transcriptase. Oncogene 23: 6890-6899.

[112] Kulawiec, M., Ayyasamy, V., Singh, K.K. (2009). p53 regulates mtDNA copy number and mitocheckpoint pathway. J Carcinog.8, 8

[113] Ahn J, Poyurovsky MV, Baptiste N, Beckerman R, Cain C, Mattia M, et al. (2009) Dissection of the sequence-specific DNA binding and exonuclease activities reveals a superactive yet apoptotically impaired mutant p53 protein. Cell Cycle 8: 1603-1615.

Chapter 4

p53 Tumor Suppressor: Functional Regulation and Role in Gene Therapy

Zeenat Farooq, Shahnawaz Wani,

Vijay Avin Balaji Ragunathrao, Rakesh Kochhar

and Mumtaz Anwar

Abstract

p53, a homo-tetrameric protein found in mammalian cells, derives its name from the fact that it settles at around 53KDa position in SDS-PAGE, due to a *"kink"* in its structure. In its functional state, p53 forms a homo-tetramer and binds to the promoters of a wide array of genes. Binding of p53 downregulates the transcription of target genes. Most of the gene targets of p53 are involved in cell cycle progression, and therefore, any malfunctions associated with p53 have catastrophic consequences for the cell. The gene encoding for p53 known as TP53 is the most well-studied gene in the entire genome because of being the most highly mutated gene in all cancer types. It is due to this widely accepted and documented *"cell protective feature"* that p53 is generally referred to as *"the guardian of the genome."* In this chapter, we will discuss the involvement of p53 in relation to carcinogenesis. We will also cover the major functions of p53 under normal conditions, major mutations of the TP53 gene, and their association with different forms of cancer.

Keywords: TP53, DNA, caspases, cell cycle, apoptosis, mutations

1. Introduction

p53 is tumor-suppressor protein also named as "the guardian of the genome" since it prevents damaged DNA from getting inserted into genome and its proliferation in daughter cells. It is p53 that decides if DNA damage is to undergo repair or cell must undergo apoptosis or enter senescence when the damage is beyond repair. If possibility of repair exists, p53 activates other genes to fix the damage; otherwise, this protein prevents the cell from dividing and signals it to undergo apoptosis. By preventing division of cells with mutated or damaged DNA, p53 helps prevent the development of tumors. Named as p53 due to its migration at 53kd size in SDS-PAGE due to its structural confirmation, 53 kDa molecular mass, this protein was identified in 1973. Its gene TP53 is located on the short arm of chromosome 17 (17p13.1) in humans [1]. Total 393 amino acids constitute p53, which are distributed into three main domains, namely transcriptional activation domain, DNA-binding domain (DBD), and tetramerization domain. Transcriptional activation domain has role to stall RNA polymerase and activate the transcriptional machinery. DNA-binding domain binds to the specific regulatory sites on the DNA response elements and is

IntechOpen

more prone to mutations due to arginine/lysine residues due to abundance of lysine / arginine residues. The tetramerization domain functions as oligomerizer domain. The binding of tetrameric p53 via DBD to regulatory DNA motifs in the genome known as response elements with the consensus sequence RRRCWWGYYYN0– 13RRRCWWGYYY (R = A or G, W = A or T, Y = C or T, N = any base) is the core event of the process. Various studies have reported dose-dependent target gene activation of p53, with high affinity of response elements for target p53 linked to cell cycle arrest and lower affinity linked to pro-apoptotic targets. This concept explains that cells undergoing feeble DNA damage are able to induce only low levels of p53, which can bind the high-affinity response elements, giving opportunity to cell for repairing its genome. However, if DNA damage is of higher level, higher p53 levels are generated to bind even the low-affinity response elements, which further activate pro-apoptotic target genes leading to cell death [2]. Mutations in p53 lead to cancers. Genome-wide analyses have shown that *TP53* is the most frequently compromised gene in human cancer [3]. Efforts to reactivate p53 function in cancer have proven to be a successful therapeutic strategy in murine models and have gained attraction with the development of a range of small molecules targeting mutant p53. Either p53 can have loss of expression during cancers or express missense mutations causing a single amino acid substitution in otherwise full-length p53 proteins. Germline p53 mutations cause Li-Fraumeni syndrome (LFS) in patients, put them at risk of different cancers at a young age. DNA-binding domain (DBD) of p53 is most prone to somatic missense mutations [4]. Cancer-derived p53 missense mutants are impaired for most wild-type (WT) p53 functions. p53 also acts as a transcription factor to control the expression of several coding and noncoding RNAs and genes including *p21, MDM2, GADD45, PERP, NOXA,* and *CYCLIN G.* In addition, p53 also suppresses the expression of some genes, such as *MAP4* and *NANOG* [5]. Also, it interacts with cytoplasmic and mitochondrial proteins to directly modulate their activity [6]. The posttranslational modifications of p53 play important roles in dictating the cellular responses to various stresses. For example, the phosphorylation of p53 at Ser46 primarily activates p53-dependent apoptosis after DNA damage. In addition, the phosphorylation of p53 at Ser315 is important for suppressing *NANOG* expression during the differentiation of ESCs. The p53 activity can also be modulated by protein–protein interaction. For example, the ASPP family proteins promote the p53-mediated apoptosis by enhancing p53-dependent induction of pro-apoptotic genes such as *PUMA* [7]. The importance of the transcriptional activity of p53 in tumor suppression is further underscored by the findings that the hotspot missense mutations of p53 in human cancers uniformly disrupt the normal DNA-binding activities of WT p53. In addition to the loss of WT p53 activity, p53 mutants also gain oncogenic activities in promoting tumorigenesis [8]. p53 has major role in detection of stress pathways, such as hypoxia and metabolic stress. In response to genotoxic and oncogenic stresses, p53 induces cell cycle arrest, apoptosis, or senescence of the stressed somatic cells to prevent the passage of the genetic abnormalities to their off springs, thus maintaining the genomic stability of mammalian cells [9]. In addition, p53 plays complex roles in cellular metabolism, contributing to p53-dependent genomic stability and tumor suppression [10]. In addition, the protein levels of p53 are also maintained at low concentration in the absence of stresses, because several E3 ligases such as MDM2 form complex with p53, leading to the ubiquitination and degradation of p53. Therefore, as potent negative regulators of p53 stability and activity, MDM2 and MDMX are oncogenes often overexpressed in human cancers to inhibit p53 function [11]. In addition to its role in cellular stability, role of p53 in embryonic stem cells (ESCs) has also been elucidated. Expansion of ESCs for dozens of passages prior to their differentiation into lineage-specific functional cells is required to harness their potential to be used in clinics for addressing different issues.

Clinical potential. Thus, its highly prevention of DNA damage and activation of oncogenic pathways are much prone to self-renewal and differentiation of ESCs. Role of p53 comes to play for maintaining the genomic stability of hESCs. However, in contrast to somatic cells, ESCs lack p53-dependent cell cycle G_1/S checkpoint, apoptosis, and senescence. Instead, when activated, p53 induces the differentiation of ESCs by directly suppressing the expression of the critical pluripotency factor Nanog. Thus, ESCs with unrepaired DNA damage or oncogenic stress are eliminated from the self-renewing pool due to the reduced Nanog expression, hence ensuring the genomic stability of self-renewing ESCs [12]. p53 is thus thought to induce the expression of the differentiation-related genes and downregulate the pluripotency genes in response to DNA damage in ESCs. In the absence of stresses, the activity of p53 must be suppressed to maintain pluripotency. The key pluripotency factor OCT4 activates the expression of histone deacetylase SIRT1, which inactivates p53 by deacetylation of p53 (13). The extensive culture of hESCs might accumulate hESCs harboring mutated p53, raising cancer risk upon long-term culture [13]. The p53 mutants might lead to *gain of functions* to promote the expression of pluripotent genes and thus the preferential expansion of hESCs harboring these p53 mutants [14]. Therefore, culture conditions that can avoid the favorable selection of hESCs harboring p53 mutations during the extended culture are required to maintain healthy ESCs. p53 also has role in inducible pluripotent stem cells (iPSCs). Biggest limitation of iPSC technology is the extremely low efficiency of successful reprogramming. p53 has been discovered to have corner stone role for reprogramming [15]. Reprogramming factors are especially c-Myc and Klf4 that are potent oncoproteins, which are often overexpressed in human cancers. The overexpression of such oncoproteins in somatic cells will activate p53, which can all block successful iPSC reprogramming and suppress the expression of Nanog that is required for maintaining pluripotency. Therefore, the silencing of the p53 gene during reprogramming has become an effective approach to increase the reprogramming efficiency [16]. In addition, proteins such as Oct4 and ZSCAN4 can promote the reprogramming efficiency by inhibiting p53 [17]. The silencing of the genes that are responsible for p53-dependent cell cycle arrest and apoptosis, such as *p21* and *Puma*, can also increase the frequency of nuclear reprogramming into induced pluripotency. On the other hand, the critical roles of p53 in maintaining genomic stability of mammalian cells raise a serious concern for the genomic instability of iPSCs as iPSCs harbor increased genetic abnormalities. First iPSC-based clinic trial to treat macular degeneration was put to halt due to high accumulation of genomic instability [13]. The genomic instability can also contribute to the immunogenicity of iPSC-derived autologous cells [18, 19]. The optimization of the reprogramming technology and the culture conditions of PSCs is required to improve PSC-based human cell therapy.

2. Stability of p53

p53 functions as a "genomic guardian" that regulates downstream targets responsible for cell fate control. p53 prevents various types of stresses such as DNA damage, hypoxia, metabolic stress, from expressing their consequences on genome, and progeny of new cells [20, 21]. The activity of p53 is tightly regulated by a complex network that includes an abundance of stress signals, posttranslational modifications, and various signaling pathways. Under normal cellular physiology, p53 is a short-lived protein and expressed at very low levels. Under stressful conditions, p53 is accumulated in the cell, and its degradation is prevented [22]. The stability of p53 is controlled predominantly by several E3 ligases, including the major ligase MDM2-mediated ubiquitination and subsequent proteasome-dependent degradation by the 26S proteasome [23, 24]. p53 levels remain low in normal non-stimulated cells. MDM2 maintains level of p53, by promoting the polyubiquitination of p53 and its degradation by proteasome pathway [25]. The major ubiquitination sites of p53 mediated by MDM2 are six lysine residues at the carboxy terminus (K370, K372, K373, K381, K382, and K386) [26]. Further, p53 is negative regulator of the MDM2. Increased p53 levels can induce MDM2 expression, leading to a decrease in p53 expression [25, 26]. MDM4 is similar to MDM2 and inhibits p53-mediated transactivation. Inhibition of MDM2 and MDM4 causes accumulation of p53 and its activity. p53 expression is also induced upon its release from its negative regulatory factors. Ubiquitination is another mechanism to prevent p53 from binding to the downstream targets, leading to apoptosis and cell cycle arrest [27, 28]. Ubiquitinylation agents of p53 have been identified, such as Pirh2, ICP0, COP1, TOPORS, ARF-BP1, CHIP, Ubc13, synoviolin, EF41, CARP2, WWP1, MSL2, E6-AP, TRIM2454, and MKRN1 [29]. Ubc13, WWP1, E4F1, and MSL2 are E3 ligases. Besides these E3 ligases, MDM2 at low level also mediates mono-ubiquitination of p53, causing proteasome-independent p53 ubiquitination [30]. Type of ubiquitination on p53 determines its effects on p53 function. E3 ligases and MDM2 can mediate lysine-48-linked polyubiquitination of p53 and target it to the 26S proteasome for degradation. Other types of ubiquitination, including mono- or lysine 63-linked polyubiquitinations, regulate nuclear export and cytosolic localizations of p53 [31].

Upon DNA damage, the interaction between p53 and MDM2 is suppressed, resulting in increasing levels of p53 protein and transcriptional activation of p53 target genes [32]. Stress signals activate ATM kinase and the DNA-PK. These kinases are starters of signal transduction cascades, which phosphorylate the N-terminus of p53 and the C-terminus of MDM2. They further dephosphorylate the central domain of MDM2, leading to weakening the interaction of p53 and MDM2 [33–35]. This prevents degradation of p53 and its accumulation to act on stress-induced damage. Oncogenes such as Myc or Ras also act as signals to stabilize p53, but they use a different route. They induce expression of p14/16^{ARF}, which binds to MDM2, inhibits its ubiquitin ligase activity, sequesters MDM2 in the nucleolus, and promotes MDM2 degradation [36, 37].

However, proteasomal degradation of p53 also has been shown to occur independently of MDM2 if newly synthesized p53 is being intrinsically unstructured [38–40]. The mechanism behind the MDM2-independent system involves Isg15-modifying system. The system is associated with the translational machinery and targeting of newly synthesized proteins [41, 42]. Different types of stimuli induce Isg15, and these include type 1 IFNs, lipopolysaccharide, and viruses [43]. Identified as a p53 target, Isg15 is also induced during the chemotherapy and requires functional p53. ISGylation is a process similar to ubiquitinylation wherein conjugation to the target proteins occurs in a three-step cascade mechanism. UBE1L is Isg15-activating E1, UBCH8 is E2 Isg15-conjugating enzyme and Isg15 E3 ligase with HERC5 being the main E3 ligase for Isg15. ISGylation negatively regulates the ubiquitin-proteasome pathway by direct interference with polyubiquitination, providing evidence of potential cross talk between these two systems [44]. It has been found that p53 is efficiently ISGylated by HERC5 and subsequently degraded by the 20S proteasome. Furthermore, Isg15 deletion increases the misfolded, dominant-negative p53, so it has been proposed that ISGylation is likely to work as a signal for degradation of misfolded p53, and this regulation is important for p53-mediated biological function [45].

The stability of both wild-type p53 and mutant p53 has been shown to be regulated by lipid messenger phosphatidylinositol 4,5-bisphosphate (PI4,5P$_2$)

$PI4,5P_2$ directly binds to protein targets known as $PI4,5P_2$ effectors and regulates their function by modulating activity and localization. The majority of $PI4,5P_2$ is generated by phosphorylation of PI4P and PI5P by type I and type II phosphatidylinositol phosphate kinases (PIPKs), respectively, and each type has α, β, and γ isoforms in humans. $PI4,5P_2$ is also found in the nucleus. Nuclear $PI4,5P_2$ is distinct from the nuclear envelope and is found in non-membranous structures such as nuclear speckles [46–48]. p53 associates with $PI4,5P_2$-generating enzyme, type Iα phosphatidylinositol-4-phosphate 5-kinase (PIPKIα) in the nucleus, and of PIPKIα diminishes p53 stability. Moreover, $PI4,5P_2$ generated by PIPKIα interacts to p53 to promote binding of HSP27 and αB-Crystallin. Both $PI4,5P_2$ binding and recruitment of HSP27 are required for stabilization of nuclear p53. Thus, PIPKIα and the PIPKIα-p53-$PI4,5P_2$-sHSP complex have been reported as promising therapeutic targets in cancer [49].

Notably, other posttranslational modifications, such as acetylation, methylation, neddylation, and sumoylation, play important roles in regulating p53 transcriptional activities. p53 is among the first non-histone proteins known to be regulated by acetylation and deacetylation [50, 51]. There are more than 50 sites in p53, which are regulated through posttranslational modifications such as phosphorylation, acetylation, methylation, and so on. These modifications have been shown to play role in regulating the stability of p53. Phosphorylation of p53 at serine and threonine residues of its N and C terminal regions takes place as a result of cell stimulation. Some of the phosphorylation sites, however, are phosphorylated in unstimulated cells and become dephosphorylated as a result of DNA damage [52]. The quintessential phosphorylation in p53 takes place at serine 15 residue and induces its dissociation from MDM2, resulting in its stability and activation of downstream functions, whereas phosphorylation at position 392, induced by DNA damage, plays a role in activation of sequence-specific DNA-binding property of p53. Phosphorylation also plays a role in formation of functionally active tetramers of p53. The transactivation domain (TAD) of p53 forms two domains TAD1 and TAD2. TAD2 interacts with the p62 family of transcription factors, which initiate chromatin decondensation at promoters. Similar to phosphorylation, acetylation of lysine residues of p53 plays role in a variety of functions through its stabilization and activation. Acetylation of p53 inhibits cell cycle progression at G2 phase and SIRT1deacetylase interacts with p53 in the nucleus, specifically deacetylating the K382 acetylation of p53 [53]. Different stimuli induce p300-mediated acetylation of lysine residue 305, both *in vitro* and *in vivo*. Lysine K320 acetylation plays role in regulation of p53 shuttle between nucleus and cytoplasm. It is also involved in BAX-mediated apoptosis after DNA damage in intestinal adenomas. In addition to acetylation, methylation of lysine and arginine also regulates p53 function. For example, K372 methylation enhances p53 stability, increases its binding to chromatin, and promotes transcriptional activity, whereas K370 methylation inhibits transactivation of p53. Additionally, arginine methylation also acts as an important regulatory mechanism for modulating p53 activity. As a result of DNA damage, p300 recruits arginine methyltransferase PRMT5 to p53, which helps in the oligomerization of p53 to modulate its transcriptional activity, whereas lack of PRMT5 alters specificity of p53 binding and triggers apoptosis. Also, siRNA-mediated knockdown of PRMT5 reduces the protein levels of p21, one of the downstream targets of p53. Mutations in the arginine residue Arg 337 are also known to be related to development of tumor and changes in biochemical characteristics of p53 oligomers. Further studies into the role of arginine mutations and enzymes, which methylate these residues (arginine methyltransferases), can lead to exploration of novel mechanisms of p53 regulation and function [54].

3. Role of p53 in carcinogenesis

p53 is involved in mitigating cellular stress such as hypoxia, DNA damage, and oncogene activation by initiating stress response mechanisms that play role in preserving genome integrity. p53 protein is famously known as the "tumor suppressor p53" because the normal functioning of p53 acts as a huge roadblock to cancer initiation and progression [55, 56]. Therefore, for carcinogenesis to take place, mutations in the p53 gene TP53 are required, which can have a significant impact on the function of p53. This is part of the reason that p53 is one of the most frequently mutated proteins in all cancers, with the numbers being as high as 53% frequency of p53 mutations in all cancers. We will discuss some of the most important p53 mutations and their effects on carcinogenesis in the next section.

3.1 Loss of p53 function

p53 is known to act as a transcription factor by binding to various DNA response elements in the target sequences. More than 100 response genes of p53 have been identified, which include CDKN1A (p21 encoding gene), BBC3, PERP, and BAX (apoptosis genes), THBS1 (angiogenesis gene), and so on [57]. Of all the domains of p53 protein, DNA-binding domain is very critical in mediating its interaction with response elements. Therefore, base mutations in sequence of DBD (mis-sense mutations) are linked with tumorigenesis as they lose the ability of interacting with DNA elements involved in tumor progression such as proto-oncogenes. These mutations arise in somatic cells either spontaneously or secondary to DNA damage. However, not all mutations affect the function of p53 in a similar manner, and the extent to which a mutation can affect tumor progression depends upon the residue being mutated and the region of gene carrying this mutation. The sequence of response elements within the target genes also determines level of p53 binding. Some of the tumor types carry mutations, which lead to a gain of function of p53 agonists such as Mdm2 and Mdm4. This results in suppression of p53 activity irrespective of the availability of normal levels of wild-type p53 in the cell. Mdm2 is a ubiquitin ligase, which binds to p53 and targets it for proteasomal degradation under normal conditions to maintain p53 at a low level in normal cells. MDM2 also binds to p53 mRNA to regulate its translation [58]. On the other hand, MDM2 itself is induced by p53, and therefore, the two proteins regulate each other through a negative feedback system. Upon receiving a stress stimulus such as DNA damage, p53 is post translationally modified by a variety of upstream effectors. These modifications inhibit the association of p53 and MDM2 to allow p53 binding to DNA response elements to initiate a stress response pathway. P53 response genes include cell cycle control genes, apoptotic genes, cellular senescence genes, and others. p53 mutations also arise in germline cells in individuals with Li Fraumeni syndrome and lead to an increased risk of developing adrenocortical, brain, and breast tumors [59].

3.2 Mutations in p53

Most of the mutations in TP53 are intronic, with no established role in tumorigenesis. Only 19 of these mutations are exonic among which 11 are nonsynonymous (replacement of one amino acid with another as a result of base change) and four are synonymous (replacement of a codon with another coding for the same amino acid). Molecular evidence suggests that P47S [60] and R72P [61] mutations lead to changes in p53 binding to response elements. Polymorphisms also exist in the response element sequences of p53 target gene promoters, which can alter the binding of p53 up to 1000-fold [62]. p53 mutations have also been identified in 50%

of adult neoplasia including the colon, lung, esophagus, stomach, liver, breast, and uterine cervix; however, no molecular data are available so far to explain the mechanism behind these mutations [63]. Also, p53 mutations occur more frequently in carcinomas than adenomas, suggesting that these represent a late event in clinical carcinogenesis. Some synonymous mutations have also been shown to alter binding of p53 mRNA to MDM2.

3.3 p53 protein: Protein interactions and carcinogenesis

Apart from DNA binding and transcriptional control, p53 also binds directly to various proteins to exhibit its tumor suppressor activity. These include cell cycle control, DNA repair, and apoptotic genes [64]. It binds to Bcl-2 family of proteins (pro-apoptotic proteins) in cells with damaged DNA to release intermembrane molecules of mitochondria and triggers apoptosis [65]. Tumor-associated mutations in p53 also affect its protein–protein interactions to promote carcinogenesis.

3.4 Role of posttranslational modifications of p53 in carcinogenesis

p53 can be modified by a variety of posttranslational modifications such as phosphorylation, acetylation, methylation, and ubiquitination on multiple residues [66, 67]. These modifications could alter the ability of p53 binding to response elements, protein–protein interactions as well as stability [68]. The first step toward p53 activation is phosphorylation of p53 at residues S10, S20, and T18 by a range of upstream kinases such as ATM and DNA-PK to increase its stability by abrogating its interaction with Mdm2. The choice of the residue targeted for phosphorylation depends on the upstream kinase and the pathway being activated. Activated p53 acts as a transcription factor by binding as a tetramer and phosphorylation at S392 at its C-terminal enhances the stability of p53 tetramer. Epigenetic modifications are also known to regulate p53 activity with both activating and repressive effects. Increase in methylation of p53 promoter decreases its rate of transcription [69] while acetylation of p53 by CBP/p300 increases its activity by inhibiting its binding with MDM2 [70].

4. p53 as tumor suppressor and DNA damage sensor

In the year 1989, the independent studies led by Bert Vogelstein and Joh Minna for the first time reported the presence of p53 mutations in colorectal and lung cancer cells [71, 72]. These studies emphasized that the genetic abnormalities of p53 show gross changes such as homozygous deletion and abnormally sized messenger RNAs along with a variety of point or small mutations and change of amino acid sequence in the region highly conserved between mouse and human. These studies were subsequently confirmed by other groups stating the importance of p53 gene in various cancer types [73–77]. However, the tumor suppressor function of p53 was first confirmed by Stephen Friend group in 1990, in which, they demonstrated the existence of somatic and germline p53 mutations in families with Li-Fraumeni syndrome, in which affected members are genetically predisposed to cancer. In all the families studied, there was a close correlation between transmission of the mutant allele and development of cancer. There are currently more than 55,000 literature reports in various types of human cancer [78–80].

The p53 tumor functions are influenced by several factors such as cell type, tissue microenvironment, and oncogenic events acquired during the tumor initiation. The activation of p53 can occur in response to DNA damage, oncogene activation, and hypoxia, in which p53 subsequently orchestrates the biological outputs such as cell-cycle arrest, senescence, apoptosis, and autophagy modulation (**Figure 1**).

4.1 The tumor suppressor function of p53 in response to cellular stress comprises three basic steps

4.1.1 Stabilization of p53

Mouse double minute2 homolog (MDM2) plays an important role in negatively regulating p53 function. Hence, the initial stabilization phase of p53 attained through actions that disrupt its interaction with MDM2. Posttranslational modification of p53 such as the amino-terminal phosphorylation by various cytoplasmic kinases prevents the binding of MDM2, which results in stabilization of p53 in response to DNA damage from ionizing radiation or certain chemotherapeutic agents (**Figure 1**) [81, 82].

4.1.2 Sequence-specific DNA binding

Once p53 is stabilized, it will bind to DNA in a sequence-specific manner. The p53 protein consists of a carboxy-terminal basic DNA-binding domain, and majority of the tumor-associated mutations in p53 protein occur in this domain, hence it is the "hot spot" for mutations. The ubiquitous DNA-binding activity of carboxy-terminal domain of p53 is to assist DNA binding and the search of p53 target sites subsequent to cellular stress [83–85].

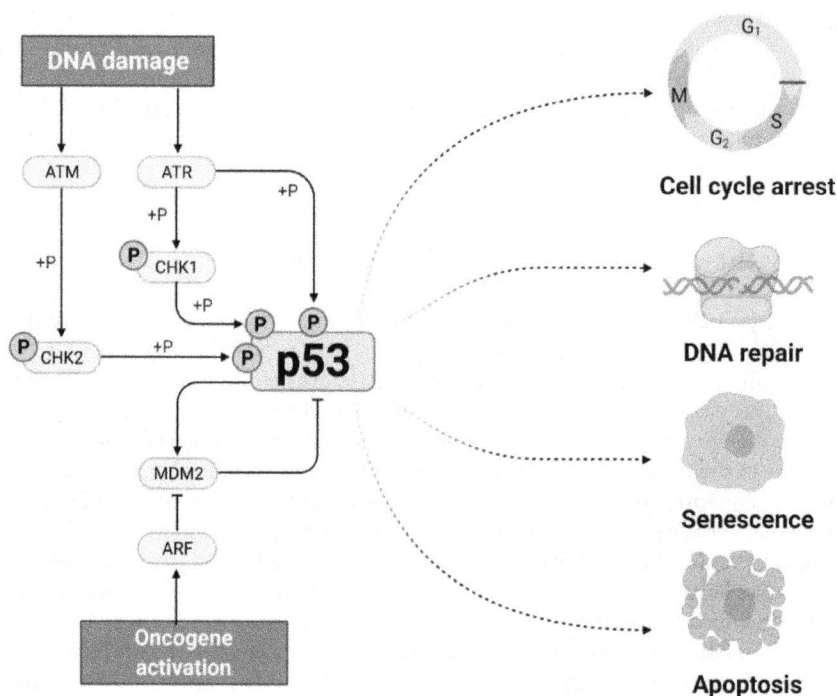

Figure 1.
Role of p53 in tumor suppressor function. The DNA damage signal sensed by ATM/ATR and oncogene activation leading to MDM2 inhibition resulting in the activation of secondary sensor p53 to orchestrates, cell cycle arrest/DNA repair/senescence/apoptosis.

4.1.3 Transcriptional activation of target genes

After stabilization of p53 and sequence-specific DNA binding, p53 activates to repress its target genes. p53 promotes transcriptional activation or repression of target genes by interacting with general transcription factors such as Transcription factor II D (TFIID) or TBP-associated factors (TAFs) depending on the complexity of promotor selection. Recent studies have reported that posttranslational modifications of p53 can influence the recruitment of p53-binding proteins to specific promoters. The interaction of CBP/p300 with p53 results in the posttranslational modifications such as p53 acetylation along with histone acetylation leading to more open chromatin conformation near p53 targets and more active p53 protein [86–88].

4.2 p53 as a DNA damage sensor

Internal cellular responses and signal transduction to genotoxic stress resulting in the activation of various transcription factors are a very complex process starting with "sensing" of the DNA damage [88]. There have been extensive studies across various disciplines exploring the activity of p53 in sensing the DNA damage induced during genotoxic stress [89]. Identifying the specific residues modified on p53 in response to DNA damage has allowed a greater understanding of the molecules that may signal to p53. Initial studies showed that the DNA damage response was sensed by p53 via Phosphoinositide 3 kinase (PI-3k) and PI 3 kinase like family members as being instrumental in mediating phosphorylation of serine-15 and regulating p53 in response to DNA damage [90–92]. Despite mediating intracellular singling events by way of phosphorylating inositol lipids, PI-3k family members have recently been expanded by the identification of their role in phosphorylating p53 to sense DNA damage. Members of this subfamily include the catalytic subunit of DNA-dependent protein kinase catalytic subunit (DNA-PKcs), Ataxia-telangiectasia mutated (ATM), ATM Rad3-related (ATR), and Transformation/Transcription domain-associated protein (TRRAP) [93–100]. Overall, p53 also functions as sensor of DNA damage by working with aforementioned kinase family members.

5. Role of p53 in apoptosis and cellular stress

During homeostatic condition, the nontransformed cells express very low or often undetectable amount of p53 protein, whereas it may still show readily detectable mRNA expression. In naïve cells, the level of p53 protein is very unstable with a half-life ranging from 5 to 30 min, owing to continuous degradation largely mediated by MDM2. The MDM2 was identified as oncogene formed as a complex with p53 protein for the first time in 1992 [101]. Ever since, persuasive evidence has emerged for MDM2 to have a physiologically critical role in controlling p53. Many findings emphasized that MDM2 itself is the product of a p53-inducible gene [25, 102, 103]. Thus, both MDM2 and p53 are linked to each other through an autoregulatory negative feedback loop intended at maintaining low cellular p53 levels under naïve condition. MDM2 is known to harbor a p53-specific E3 ubiquitin ligase activity within its evolutionarily conserved Zinc-binding domain, which is critical for its E3 ligase activity [104]. Studies have shown that MDM2 is largely expressed in nucleus and bound to p300/CBP leading to the p53 ubiquitination [105]. The crystallographic studies showed the biochemical basis of MDM2-mediated inhibition of p53 function, in which, the amino terminal domain of MDM2 forms a deep hydrophobic cleft into which transactivation domain p53 binds, thereby concealing itself from interaction with the transcriptional machinery [106].

During cell transformation, caused due to stress stimuli such as activation of oncogenes or any other DNA damage signals, the p53 protein level substantially increases because of the activation of survival pathways, which lead to the inhibition of MDM2 and posttranslational modifications in the p53 protein itself. The activated p53 thus turns on the diverse cellular effector process, including cell cycle arrest, cellular senescence, DNA damage-repair pathways, and apoptotic cell death [25, 107–111].

In 1991, studies from Moshe Oren group showed for the first time that p53 can induce apoptosis of myeloid leukemia cells. They showed that temperature-sensitive conditionally active mutant of p53, in which at 37°C behaves as mutant but at 32°C it assumes wild-type (WT) p53 structure and function and starts inducing apoptosis of leukemia cells *in vitro* [112]. Further these studies using temperature-sensitive p53 or WT p53 were confirmed by various other groups in different cancer cell lines such as erythroleukemia, colon cancer, and Burkitt lymphoma [113, 114].

5.1 P53-mediated intrinsic apoptotic pathways

Broadly, the mammalian cells endure apoptosis in two distinct manners. Intrinsic or mitochondrial stress apoptotic pathway, activated during stress conditions, such as cytokine deprivation, ER stress, or DNA damage, which is regulated by B-Cell Lymphoma 2(BCL-2) family proteins. Another mechanism of apoptotic activation of mammalian cells due to ligation of members of tumor necrosis factor receptor (TNFR) family bearing intracellular death domain, called as extrinsic apoptotic pathway [115].

The intrinsic apoptotic pathway or mitochondrial pathway is initiated by the release of apoptogenic factors such as cytochrome *c*, apoptosis-inducing factor (AIF), Smac (second mitochondria-derived activator of caspase)/DIABLO (direct inhibitor of apoptosis protein (IAP)-binding protein with low PI), Omi/HtrA2 or endonuclease G from the mitochondrial intermembrane space. The release of cytochrome *c* into the cytosol triggers caspase-3 activation through formation of the cytochrome *c*/Apaf-1/caspase-9-containing apoptosome complex, whereas Smac/DIABLO and Omi/HtrA2 promote caspase activation through neutralizing the inhibitory effects to the IAPs (**Figure 2**) [116, 117].

In vitro studies using overexpression of WT p53 or temperature-sensitive p53 showed that elevated expression of anti-apoptotic BCL-2 could prevent p53 induced apoptosis. Further, the cells rescued from p53-induced apoptosis by elevated expression of BCL-2 still able to perform cell cycle arrest, indicating BCL-2 does not directly block p53, hence p53 was fully functional. Thus, BCL-2 and its family member proteins inhibit the p53-induced apoptosis at a downstream point of apoptosis signaling, suggesting induction of cell cycle arrest and apoptosis by p53 through distinct pathways [118, 119]. Though, the study using overexpressed WT and temperature-sensitive p53 mutants provides efficient evidence for apoptosis induction, it does not resemble the physiological condition where p53 protein level is normal [111]. All these caveats are addressed once after the generation of p53 knockout animals, in which, the thymocytes and other lymphoid cell subsets are completely resistant to apoptosis induced by γ-radiation and treatment with chemotherapeutic drugs that induce DNA damage.

The most intuitive link between p53 and BCL-2, unveiled the quest to identify the p53 activated initiators of the cell death pathways that is regulated by BCL-2. Many downstream effectors were identified in response to overexpression of p53 at both physiological and nonphysiological level. These include BCL-2 associated X protein (Bax), BCL-2 homolog 3 (BH3)—only proteins, p53 upregulated modulator of apoptosis (PUMA), and BH3 interacting domain death agonist (Bid). Gene targeting

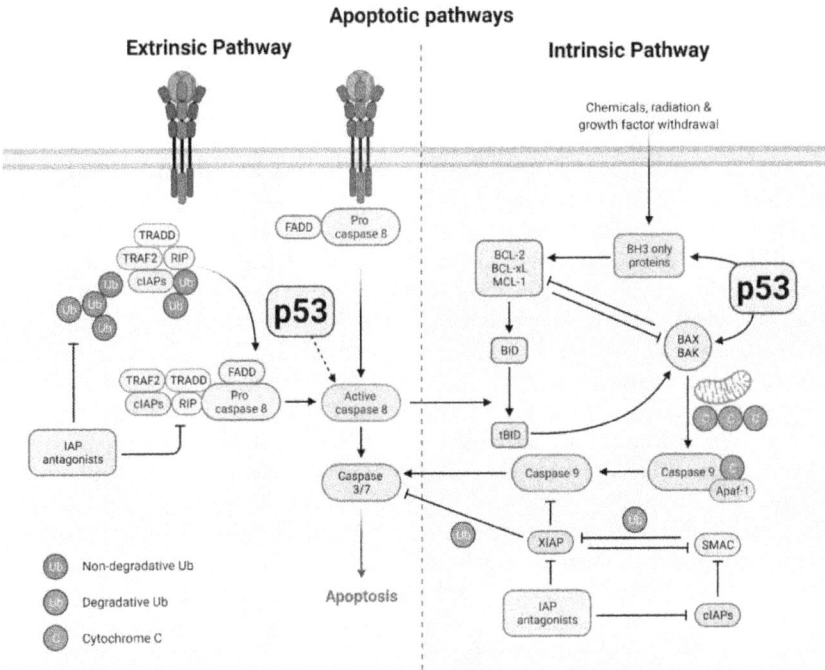

Figure 2.
P53-mediated apoptotic pathways and their cross talk. Extrinsic apoptotic pathways mediated by p53 via activating caspase 8 leading to the activation of Caspase3 and 7. Intrinsic apoptotic pathway mediated by p53 leading to the activation of pro-apoptotic molecules BH3 only proteins and BAX/BAK leading to the activation of caspase 9 via mitochondrial mediated cytochrome c, further activating caspase 3 and 7 to induce apoptosis.

studies in both *in vivo* and *in vitro* showed that pro-apoptotic members of BCL-2 family can act downstream of p53 during apoptosis. The studies using Bax-knockout mouse embryo fibroblast showed that they are desensitized to oncogene-induced and p53-dependent apoptosis leading to suppression of tumorigenesis [120]. Some *in vitro* studies showed that knocking down either Bax or PUMA in cell lines induces various levels of apoptotic defects [121–125]. Taken together, these studies suggest that loss or mutation of p53 attenuates the expression of the downstream targets implying that the phenotypes of the attenuated effectors show defects in p53-mediated apoptosis.

Though, p53-mediated intrinsic pathway of apoptosis controls the factors that act upstream of the mitochondria, it can also transactivate several components of the apoptotic effector machinery [124]. Apoptosis protease activating factor 1 (Apaf-1), caspase-6, and E2 factor family transcription factor (E2F) are the apoptotic effectors regulated by p53. Apaf-1 is known to act as coactivator of caspase-9 and helps initiate caspase cascade, and p53 loss can interfere in Apaf-1-mediated caspase cascade initiation [126]. In addition, p53 can upregulate the caspase 6, which is known as an effector caspase, leading to enhanced chemosensitivity of some cell types [127]. Likewise, p53 interferes with E2F and could result in promoting apoptosis and increase the caspase expression through a direct transcriptional mechanism [126].

5.2 p53-mediated extrinsic apoptotic pathways

Extrinsic apoptotic pathway also known as death receptor-mediated apoptotic pathway occurs under stress condition due to stimulation of death receptors of the tumor necrosis factor (TNF) receptor superfamily such as CD95 (APO-1/Fas) or TNF-related apoptosis-inducing ligand (TRAIL) receptors result in activation of the initiator caspase-8, which can propagate the apoptosis signal by direct cleavage of downstream effector caspases such as caspase-3 (**Figure 2**).

Several studies support the hypothesis of p53-mediated intrinsic pathway of apoptosis, few also showed that the extrinsic apoptotic pathway can also regulated by p53, although the overall contribution of p53-mediated extrinsic apoptotic pathway remains debatable and is still being researched [124]. Some of the proteins that are involved in extrinsic mechanisms such as Fas/CD95, Fas ligand, and death receptor 5 (DR5) are shown to be direct targets of p53 [128–130]. Moreover, some studies have also shown that there is a cross talk between intrinsic and extrinsic pathways because of the ability of p53 to transactivate Bid [125]. Consequently, p53 may sensitize cells to death receptor ligands, either inducing apoptosis directly or enhancing cell death in ligand-rich environment. Some studies have shown that disabling p53 sensitization to death receptor ligands by mutating p53 can promote drug resistance and can provide an ambient tumor microenvironment with immune privilege.

Despite being directly involved in both intrinsic and extrinsic apoptotic pathways, p53 can also regulate the survival signals indirectly. Phosphatase tensin homolog at chromosome 10 (PTEN) is a lipid phosphatase, known to inhibit phosphoinositide 3-kinase induced survival signaling by dephosphorylating 3'-phosphorylated phosphatidylinositides (PIP3) to 2'-phosphorylated phosphatidylinositides (PIP2). Some studies suggest that p53 can interfere in survival signaling by transactivating the PTEN promotor leading to increased expression of PTEN. Although, the disruption of PTEN can compromise p53-mediated apoptosis [131, 132]. Independent studies from Puzio-Kuter et al., and Freeman et al., showed that the tumor suppressor PTEN regulates the activity of p53 and levels of p53 levels through mechanisms involving both phosphatase dependent and independent manner [131, 132]. Thus, in the case of PTEN mutation or loss, p53 can neutralize the survival signals, seemingly diminishing the threshold needed for proapoptotic factors to trigger cell death.

6. Role of p53 in gene therapy

Most of the human cancer types show altered p53 level, and hence, the concept of restoration of p53 for cancer therapy is a very attractive strategy. The normal function of mutated p53 can be restored using various pharmacologically active small molecular inhibitors, by inducing massive apoptosis or reactivating p53. The compounds such as PRIMA-1, CP-31398, and SH group targeting compounds induce the apoptosis and reactivate the p53 [133, 134]. Another set of small molecular inhibitors such as Nutlin-3a, RG7112, CGM097, and SAR405838 block the interaction of p53 with MDM2 [135–137]. Though, the abovementioned small molecular inhibitors have shown good anticancer efficacy, they also have some limitations. For instance, it is still unclear whether PRIMA-1 and similar compounds effectively target all mutant p53 variants and whether the tumor suppressor functions of p63 and p73 could be negatively affected [133]. Similarly, the compounds inhibiting p53 and MDM2 interactions are not shown to be effective in tumors with a high prevalence of p53 mutations. All these impediments with p53 target using pharmaceutical agents paved a new way toward gene therapy. Scientists around the globe developed various techniques in the field of biotechnology to deliver the healthy, normal functioning p53 gene to cells that turned into cancerous due to the mutations in the p53 gene. Different mode of p53 gene therapy using different vectors is discussed below.

6.1 Nonreplicating viruses-based p53 gene therapy

This process involves a healthy p53 genes and a vehicle viral vector, in which the viral DNA has been altered to prevent it from replicating. This "safe version" of viral vector is then used to transport healthy p53 into transformed cells by directly injecting into the tumor site. If the transduction is successful, the p53 gene will make a functional p53 protein within the tumor microenvironment apparently restoring the normal p53 cellular function and thereby preventing cancer growth [138]. The history of p53 gene therapy is quite interesting. In 1992, Dr. Jack A Roth from MD Anderson Cancer center led a team to investigate the first p53 gene therapy clinical trials, which were approved by the National Institute of Health (NIH) and US Food and Drug Administration (FDA). He demonstrated and proved the gene therapy efficacy of p53 through laboratory and preclinical studies, which led to the approval for this historic protocol. In brief, they developed a retroviral and adenoviral vector expressing p53 tumor suppressor gene and completed the first clinical trials in lung cancer patients against non-small-cell lung carcinoma, by showing that restoration of function for a single tumor suppressor gene could mediate regression of human cancer in vivo [138–141]. Adenovirus p53 became the first gene therapy approved for human use. Ever since, thousands of patients received different p53 mediated gene therapy with replication deficient viral vectors under many clinical trials without any significant adverse effect.

6.2 Oncolytic virus-based p53 gene therapy

The oncolytic virus therapy utilizes replication-competent viruses to kill malignant cells, leaving normal cells unscathed. Studies have shown limited success using a modified form of the measles virus to target the mesothelioma cells using oncolytic viruses alone [142]. Researchers hope that combining p53 gene with oncolytic viruses to make the treatment more effective than gene therapy or oncolytic therapy alone and these studies are still under preclinical trial stage. The preclinical studies combining p53 gene therapy along with oncolytic viruses suggested that the expression of WT p53 transgenes improves the oncolytic virus therapy safety, onco-selectivity, increases onco-toxicity, and augments antitumor effects by promoting the stimulation of anticancer immune responses [143].

6.3 Nanoparticles-based p53 gene therapy

Another easy and convenient method of delivering p53 gene is using synthetic nanoparticles. Like viruses, nanoparticles could be designed to deliver their contents to cancer cells specifically, leaving healthy cells unaffected. Nanoparticle-based p53 gene therapy is still under clinical trials, and these studies suggested that the nanoparticle tagged with p53 gene would be a safer mode of gene delivery method compared with viruses, creating no risk of infection, and they could travel through the body without provoking a response from the immune system [144–146].

7. Conclusion

Most of the "tumor associated mutations" in p53 are single base substitutions in the coding sequence [147]. Apart from these, more than 200 single nucleotide polymorphisms have been identified in TP53 with no measurable consequence on p53 function and/or tumor progression. p53 is unique for being the most well-studied and most frequently mutated tumor suppressor gene with a wide spectrum of residual activity as a direct consequence of the mutated residue [148]. The molecular basis behind most of the p53 mutations is not well understood. Therefore, better understanding of these mechanisms could lead to improvements in clinical treatment of cancers carrying p53 mutations. Recently, other mechanisms such as micro RNAs have been implicated to be involved in p53-mediated gene regulation and which further necessitates undertaking more studies to understand the roles of the *"guardian of the genome"* in a more elaborate manner. Because of the diversity of mutations TP53 can carry a large number of online resources are available that contain information on TP53 mutations, domain containing the mutated residue, approximate loss of function, and possible association with cancer types. Some of the most well-known include the IARC TP53 mutation database, the p53 Knowledgebase, the TP53 Web Site, and the Database of germline p53 mutations.

Acknowledgements

Figures are made using Biorender. Zeenat Farooq is highly acknowledged for fine editing.

Author details

Zeenat Farooq[1†], Shahnawaz Wani[2†], Vijay Avin Balaji Ragunathrao[1*], Rakesh Kochhar[3*] and Mumtaz Anwar[1*]

1 Department of Pharmacology and Regenerative Medicine, College of Medicine, University of Illinois at Chicago, Chicago, USA

2 Department of Medicine, SKIMS, Srinagar, Jammu & Kashmir, India

3 Department of Gastroenterology, Postgraduate Institute of Medical Education and Research, Chandigarh, India

*Address all correspondence to: avin01@uic.edu; dr_kochhar@hotmail.com and mumtazan@uic.edu

† These authors contributed equally to this work.

IntechOpen

References

[1] Kamada R, Toguchi Y, Nomura T, Imagawa T, Sakaguchi K. Tetramer formation of tumor suppressor protein p53: Structure, function, and applications. Biopolymers. 4 Nov 2016;**106**(4):598-612. DOI: 10.1002/bip.22772. PMID: 26572807

[2] Farkas M, Hashimoto H, Bi Y, et al. Distinct mechanisms control genome recognition by p53 at its target genes linked to different cell fates. Nature Communications. 2021;**12**:484

[3] Kandoth C, Mclellan MD, Vandin F, Ye K, Niu B, Lu C, et al. Mutational landscape and significance across 12 major cancer types. Nature. 2013;**502**:333-339. DOI: 10.1038/nature12634

[4] Bougeard G, Renaux-Petel M, Flaman JM, Charbonnier C, Fermey P, Belotti M, et al. Revisiting Li-Fraumeni syndrome from *TP53* mutation carriers. Journal of Clinical Oncology. 2015;**33**

[5] Sullivan KD, Galbraith MD, Andrysik Z, Espinosa JM. Mechanisms of transcriptional regulation by p53. Cell Death and Differentiation. 2018;**25**:133-143

[6] Ho T, Tan BX, Lane D. How the other half lives: What p53 does when it is not being a transcription factor. International Journal of Molecular Sciences. 2020;**21**:13

[7] Trigiante G, Lu X. ASPP [corrected] and cancer. Nature Reviews. Cancer. 2006;**6**:217-226

[8] Sabapathy K, Lane DP. Therapeutic targeting of p53: All mutants are equal, but some mutants are more equal than others. Nature Reviews. Clinical Oncology. 2017;**15**:13

[9] Eischen CM. Genome stability requires p53. Cold Spring Harbor Perspectives in Medicine. 2016;**6**:a026096

[10] Li L, Mao Y, Zhao L, Li L, Wu J, Zhao M, et al. p53 regulation of ammonia metabolism through urea cycle controls polyamine biosynthesis. Nature. 2019;**567**:253-256

[11] Oliner JD, Saiki AY, Caenepeel S. The role of MDM2 amplification and overexpression in tumorigenesis. Cold Spring Harbor Perspectives in Medicine. 2016;**6**:a026336

[12] Lin T, Chao C, Saito SI, Mazur SJ, Murphy ME, Appella E, et al. p53 induces differentiation of mouse embryonic stem cells by suppressing Nanog expression. Nature Cell Biology. 2005;**7**:165-171

[13] Zhang Z-N, Chung S-K, Xu Z, Xu Y. Oct4 maintains the pluripotency of human embryonic stem cells by inactivating p53 through Sirt1-mediated deacetylation. Stem Cells. 2014;**32**:157-165

[14] Merkle FT, Ghosh S, Kamitaki N, Mitchell J, Avior Y, Mello C, et al. Human pluripotent stem cells recurrently acquire and expand dominant negative P53 mutations. Nature. 2017;**545**:229-233

[15] Koifman G, Shetzer Y, Eizenberger S, Solomon H, Rotkopf R, Molchadsky A, et al. A mutant p53-dependent embryonic stem cell gene signature is associated with augmented tumorigenesis of stem cells. Cancer Research. 2018;**78**:5833

[16] Smith ZD, Nachman I, Regev A, Meissner A. Dynamic single-cell imaging of direct reprogramming reveals an early specifying event. Nature Biotechnology. 2010;**28**:521-526

[17] Zhao T, Xu Y. p53 and stem cells: New developments and new concerns. Trends in Cell Biology. 2010;**20**:170-175

[18] Mandai M, Watanabe A, Kurimoto Y, Hirami Y, Morinaga C, Daimon T, et al.

Autologous induced stem-cell-derived retinal cells for macular degeneration. The New England Journal of Medicine. 2017;**376**:1038-1046

[19] Deuse T, Hu X, Agbor-Enoh S, Koch M, Spitzer MH, Gravina A, et al. De novo mutations in mitochondrial DNA of iPSCs produce immunogenic neoepitopes in mice and humans. Nature Biotechnology. 2019;**37**:1137-1144

[20] Kruiswijk F, Labuschagne CF, Vousden KH. p53 in survival, death and metabolic health: A lifeguard with a licence to kill. Nature Reviews. Molecular Cell Biology. 2015;**16**:393-405

[21] Bieging KT, Mello SS, Attardi LD. Unravelling mechanisms of p53-mediated tumour suppression. Nature Reviews. Cancer. 2014;**14**:359-370

[22] Hu W, Feng Z, Levine AJ. The regulation of multiple p53 stress responses is mediated through MDM2. Genes & Cancer. 2012;**3**:199-208

[23] Zhang Q, Zeng SX, Lu H. Targeting p53-MDM2-MDMX loop for cancer therapy. Sub-Cellular Biochemistry. 2014;**85**:281-319

[24] Leng RP, Lin Y, Ma W, Wu H, Lemmers B, Chung S, et al. Pirh2, a p53-induced ubiquitin-protein ligase, promotes p53 degradation. Cell. 2003;**112**:779-791

[25] Haupt Y, Maya R, Kazaz A, et al. Mdm2 promotes the rapid degradation of p53. Nature. 1997;**387**:296-299

[26] Rodriguez MS, Desterro JM, Lain S, et al. Multiple C-terminal lysine residues target p53 for ubiquitin-proteasome-mediated degradation. Molecular and Cellular Biology. 2000;**20**:8458-8467

[27] Haglund K, Dikic I. Ubiquitylation and cell signaling. The EMBO Journal. 2005;**24**:3353-3359

[28] Lee JT, Wheeler TC, Li L, et al. Ubiquitination of alpha-synuclein by Siah-1 promotes alpha-synuclein aggregation and apoptotic cell death. Human Molecular Genetics. 2008;**17**:906-917

[29] Lee EW, Lee MS, Camus S, et al. Differential regulation of p53 and p21 by MKRN1 E3 ligase controls cell cycle arrest and apoptosis. The EMBO Journal. 2009;**28**:2100-2113

[30] Li MY, Brooks CL, Wu-Baer F, et al. Mono-versus polyubiquitination: Differential control of p53 fate by Mdm2. Science. 2003;**302**:1972-1975

[31] Lee JT, Gu W. The multiple levels of regulation by p53 ubiquitination. Cell Death and Differentiation. 2010;**17**: 86-92

[32] Kaiser AM, Attardi LD. Deconstructing networks of p53-mediated tumor suppression in vivo. Cell Death and Differentiation. 2018;**25**:93-103

[33] Boehme A, Blattner C. Regulation of p53—Insights into a complex process. Critical Reviews in Biochemistry and Molecular Biology. 2009;**44**:367-392

[34] Carr MI, Jones SN. Regulation of the Mdm2-p53 signaling axis in the DNA damage response and tumorigenesis. Translational Cancer Research. 2016;**5**: 707-724. DOI: 10.21037/tcr.2016.11.75

[35] Boehme KA, Kulikov R, Blattner C. p53 stabilization in response to DNA damage requires Akt/PKB and DNA-PK. Proceedings of the National Academy of Sciences of the United States of America. 2008;**105**:7785-7790

[36] Zhang Y, Xiong Y, Yarbrough WG. ARF promotes MDM2 degradation and stabilizes p53: ARF-INK4a locus deletion impairs both the Rb and p53 tumor suppression pathways. Cell. 1998;**92**: 725-734

[37] Weber JD, Taylor LJ, Roussel MF, Sherr CJ, Bar-Sagi D. Nucleolar Arf sequesters Mdm2 and activates p53. Nature Cell Biology. 1999;**1**:20-26

[38] Asher G, Lotem J, Kama R, Sachs L, Shaul Y. NQO1 stabilizes p53 through a distinct pathway. Proceedings of the National Academy of Sciences of the United States of America. 2002;**99**:3099-3104. DOI: 10.1073/pnas.052706799

[39] Moscovitz O, Tsvetkov P, Hazan N, Michaelevski I, Keisar H, Ben-Nissan G, et al. A mutually inhibitory feedback loop between the 20S proteasome and its regulator, NQO1. Molecular Cell. 2012;**47**:76-86

[40] Tsvetkov P, Reuven N, Shaul Y. Ubiquitin-independent p53 proteasomal degradation. Cell Death and Differentiation. 2010;**17**:103-108. DOI: 10.1038/cdd.2009.67

[41] Polyak K, Xia Y, Zweier JL, Kinzler KW, Vogelstein B. A model for p53-induced apoptosis. Nature. 1997;**389**:300-305

[42] Durfee LA, Lyon N, Seo K, Huibregtse JM. The ISG15 conjugation system broadly targets newly synthesized proteins: Implications for the antiviral function of ISG15. Molecular Cell. 2010;**38**:722-732

[43] Yang P, Yu Z, Gandahi JA, Bian X, Wu L, Liu Y, et al. The identification of c-kit-positive cells in the intestine of chicken. Poultry Science. 2012;**91**:2264-2269

[44] Liu M, Hummer BT, Li X, Hassel BA. Camptothecin induces the ubiquitin-like protein, ISG15, and enhances ISG15 conjugation in response to interferon. Journal of Interferon & Cytokine Research. 2004;**24**:647-654

[45] Huang YF, Wee S, Gunaratne J, Lane DP, Bulavin DV. Isg15 controls p53 stability and functions. Cell Cycle. 2014;**13**(14):2200-2210. DOI: 10.4161/cc.29209

[46] Barlow CA, Laishram RS, Anderson RA. Nuclear phosphoinositides: A signaling enigma wrapped in a compartmental conundrum. Trends in Cell Biology. 2010;**20**:25-35

[47] Mellman DL et al. A PtdIns4,5P2-regulated nuclear poly(a) polymerase controls expression of select mRNAs. Nature. 2008;**451**:1013-1017

[48] Boronenkov IV, Loijens JC, Umeda M, Anderson RA. Phosphoinositide signaling pathways in nuclei are associated with nuclear speckles containing premRNA processing factors. Molecular Biology of the Cell. 1998;**9**:3547-3560

[49] Choi S, Chen M, Cryns VL, Anderson RA. A nuclear phosphoinositide kinase complex regulates p53. Nature Cell Biology. 2019;**21**(4):462-475. DOI: 10.1038/s41556-019-0297-2

[50] Gu W, Roeder RG. Activation of p53 sequence-specific DNA binding by acetylation of the p53 C-terminal domain. Cell. 1997;**90**:595-606

[51] Luo J, Su F, Chen D, Shiloh A, Gu W. Deacetylation of p53 modulates its effect on cell growth and apoptosis. Nature. 2000;**408**:377-381

[52] Dai C, Gu W. p53 post-translational modification: Deregulated in tumorigenesis. Trends in Molecular Medicine. 2010;**16**:528-536

[53] Yi J, Luo J. SIRT1 and p53, effect on cancer, senescence and beyond. Biochimica et Biophysica Acta. 2010;**1804**(8):1684-1689. DOI: 10.1016/j.bbapap.2010.05.002

[54] Che Z, Sun H, Yao W, Lu B, Han Q. Role of post-translational modifications

in regulation of tumor suppressor p53 function. Frontiers of Oral and Maxillofacial Medicine. 2020;**2**:1-15

[55] Vogelstein B, Lane D, Levine AJ. Surfing the p53 network. Nature. 2000;**408**(6810):307-310. DOI: 10.1038/35042675

[56] Vousden KH, Lane DP. p53 in health and disease. Nature Reviews. Molecular Cell Biology. 2007;**8**(4):275-283. DOI: 10.1038/nrm2147

[57] Xu Y. Regulation of p53 responses by post-translational modifications. Cell Death and Differentiation. 2003 Apr;**10**(4):400-403. DOI: 10.1038/sj.cdd.4401182

[58] Candeias MM, Malbert-Colas L, Powell DJ, Daskalogianni C, Maslon MM, Naski N, et al. P53 mRNA controls p53 activity by managing Mdm2 functions. Nature Cell Biology. 2008;**10**(9):1098-1105. DOI: 10.1038/ncb1770

[59] Malkin D. p53 and the Li-Fraumeni syndrome. Cancer Genetics and Cytogenetics. 1993;**66**(2):83-92. DOI: 10.1016/0165-4608(93)90233-c

[60] Felley-Bosco E, Weston A, Cawley HM, Bennett WP, Harris CC. Functional studies of a germ-line polymorphism at codon 47 within the p53 gene. American Journal of Human Genetics. 1993;**53**(3):752-759

[61] Matlashewski GJ, Tuck S, Pim D, Lamb P, Schneider J, Crawford LV. Primary structure polymorphism at amino acid residue 72 of human p53. Molecular and Cellular Biology. 1987;**7**(2):961-963. DOI: 10.1128/mcb.7.2.961-963.1987

[62] Resnick MA, Inga A. Functional mutants of the sequence-specific transcription factor p53 and implications for master genes of diversity. Proceedings of the National Academy of Sciences of the United States of America. 2003;**100**(17):9934-9939. DOI: 10.1073/pnas.1633803100

[63] Chang F, Syrjänen S, Syrjänen K. Implications of the p53 tumor-suppressor gene in clinical oncology. Journal of Clinical Oncology. 1995;**13**(4):1009-1022. DOI: 10.1200/JCO.1995.13.4.1009

[64] Braithwaite AW, Del Sal G, Lu X. Some p53-binding proteins that can function as arbiters of life and death. Cell Death and Differentiation. 2006;**13**(6):984-993. DOI: 10.1038/sj.cdd.4401924

[65] Leu JI, Dumont P, Hafey M, Murphy ME, George DL. Mitochondrial p53 activates Bak and causes disruption of a Bak-Mcl1 complex. Nature Cell Biology. 2004;**6**(5):443-450. DOI: 10.1038/ncb1123

[66] Farooq Z, Shah A, Tauseef M, Rather RA, Anwar M. Evolution of Epigenome as the Blueprint for Carcinogenesis [Online First]. London: IntechOpen; 2021. DOI: 10.5772/intechopen.97379

[67] Kruse JP, Gu W. Modes of p53 regulation. Cell. 2009;**137**(4):609-622. DOI: 10.1016/j.cell.2009.04.050

[68] Ryan KM, Phillips AC, Vousden KH. Regulation and function of the p53 tumor suppressor protein. Current Opinion in Cell Biology. 2001;**13**(3):332-337. DOI: 10.1016/s0955-0674(00)00216-7

[69] Pogribny IP, Pogribna M, Christman JK, James SJ. Single-site methylation within the p53 promoter region reduces gene expression in a reporter gene construct: Possible in vivo relevance during tumorigenesis. Cancer Research. 2000;**60**(3):588-594

[70] Li M, Luo J, Brooks CL, Gu W. Acetylation of p53 inhibits its

ubiquitination by Mdm2. The Journal of Biological Chemistry. 2002;**277**(52):50607-50611. DOI: 10.1074/jbc.C200578200

[71] Takahashi T, Nau MM, Chiba I, Birrer MJ, Rosenberg RK, Vinocour M, et al. p53: A frequent target for genetic abnormalities in lung cancer. Science. 1989;**246**(4929):491-494

[72] Nigro JM, Baker SJ, Preisinger AC, Jessup JM, Hosteller R, Cleary K, et al. Mutations in the p53 gene occur in diverse human tumour types. Nature. 1989;**342**(6250):705-708

[73] Rodrigues NR, Rowan A, Smith ME, Kerr IB, Bodmer WF, Gannon JV, et al. p53 mutations in colorectal cancer. Proceedings of the National Academy of Sciences. 1990;**87**(19):7555-7559

[74] Coles C, Condie A, Chetty U, Steel CM, Evans HJ, Prosser J. p53 mutations in breast cancer. Cancer Research. 1992;**52**(19):5291-5298

[75] Olivier M, Hollstein M, Hainaut P. TP53 mutations in human cancers: Origins, consequences, and clinical use. Cold Spring Harbor Perspectives in Biology. 2010;**2**(1):a001008

[76] Mogi A, Kuwano H. TP53 mutations in nonsmall cell lung cancer. Journal of Biomedicine and Biotechnology. 2011;**2011**:a001

[77] Zhang W, Edwards A, Flemington EK, Zhang K. Significant prognostic features and patterns of somatic TP53 mutations in human cancers. Cancer informatics. 2017;**16**:1176935117691267

[78] Malkin D, Li FP, Strong LC, Fraumeni JF, Nelson CE, Kim DH, et al. SH friend. Germ line p53 mutations in a familial syndrome of breast cancer, sarcomas, and other neoplasms. Science. 1990;**250**(4985):1233-1238

[79] Yonish-Rouach E, Resnftzky D, Lotem J, Sachs L, Kimchi A, Oren M. Wild-type p53 induces apoptosis of myeloid leukaemic cells that is inhibited by interleukin-6. Nature. 1991;**352** (6333):345-347

[80] Johnson PE, Chung ST, Benchimol S. Growth suppression of friend virus-transformed erythroleukemia cells by p53 protein is accompanied by hemoglobin production and is sensitive to erythropoietin. Molecular and Cellular Biology. 1993;**13**(3):1456-1463

[81] Chen D, Zhang Z, Li M, Wang W, Li Y, Rayburn ER, et al. Ribosomal protein S7 as a novel modulator of p53–MDM2 interaction: Binding to MDM2, stabilization of p53 protein, and activation of p53 function. Oncogene. 2007;**26**(35):5029-5037

[82] Blagosklonny MV. Loss of function and p53 protein stabilization. Oncogene. 1997;**15**(16):1889-1893

[83] Kern SE, Kinzler KW, Bruskin A, Jarosz D, Friedman P, Prives C, et al. Identification of p53 as a sequence-specific DNA-binding protein. Science. 1991;**252**(5013):1708-1711

[84] Liu Y, Kulesz-Martin M. p53 protein at the hub of cellular DNA damage response pathways through sequence-specific and non-sequence-specific DNA binding. Carcinogenesis. 2001;**22**(6):851-860

[85] Nishimura M, Arimura Y, Nozawa K, Kurumizaka H. Linker DNA and histone contributions in nucleosome binding by p53. The Journal of Biochemistry. 2020;**168**(6): 669-675

[86] Caelles C, Helmberg A, Karin M. p53-dependent apoptosis in the absence of transcriptional activation of p53-target genes. Nature. 1994;**370**(6486): 220-223

[87] Bieging KT, Attardi LD. Deconstructing p53 transcriptional networks in tumor suppression. Trends in Cell Biology. 2012;22(2):97-106

[88] Kokontis JM, Wagner AJ, O'Leary M, Liao S, Hay N. A transcriptional activation function of p53 is dispensable for and inhibitory of its apoptotic function. Oncogene. 2001;20(6):659-668

[89] Sakamuro D, Sabbatini P, White E, Prendergast GC. The polyproline region of p53 is required to activate apoptosis but not growth arrest. Oncogene. 1997;15(8):887-898. DOI: 10.1038/sj.onc.1201263

[90] Abraham AG, O'Neill E. PI3K/Akt-mediated regulation of p53 in cancer. Biochemical Society Transactions. 2014;42(4):798-803

[91] Fang L, Li G, Liu G, Lee SW, Aaronson SA. p53 induction of heparin-binding EGF-like growth factor counteracts p53 growth suppression through activation of MAPK and PI3K/Akt signaling cascades. The EMBO Journal. 2001;20(8):1931-1939

[92] Fan L, Ren C, Wang J, Wang S, Yang L, Han X, et al. The crosstalk between STAT3 and p53/RAS signaling controls cancer cell metastasis and cisplatin resistance via the slug/MAPK/PI3K/AKT-mediated regulation of EMT and autophagy. Oncogene. 2019;8:1-5

[93] Zhang J, De Toledo SM, Pandey BN, Guo G, Pain D, Li H, et al. Role of the translationally controlled tumor protein in DNA damage sensing and repair. Proceedings of the National Academy of Sciences. 2012;109(16):E926-E933

[94] Finzel A, Grybowski A, Strasen J, Cristiano E, Loewer A. Hyperactivation of ATM upon DNA-PKcs inhibition modulates p53 dynamics and cell fate in response to DNA damage. Molecular Biology of the Cell. 2016;27(15): 2360-2367

[95] Banin S, Moyal L, Shieh SY, Taya Y, Anderson CW, Chessa L, et al. Enhanced phosphorylation of p53 by ATM in response to DNA damage. Science. 1998;281(5383):1674-1677

[96] Karlseder J, Broccoli D, Dai Y, Hardy S, de Lange T. p53-and ATM-dependent apoptosis induced by telomeres lacking TRF2. Science. 1999;283(5406):1321-1325

[97] Tibbetts RS, Brumbaugh KM, Williams JM, Sarkaria JN, Cliby WA, Shieh SY, et al. A role for ATR in the DNA damage-induced phosphorylation of p53. Genes & Development. 1999;13(2):152-157

[98] Wu H, Zhou X, Wang X, Cheng W, Hu X, Wang Y, et al. miR-34a in extracellular vesicles from bone marrow mesenchymal stem cells reduces rheumatoid arthritis inflammation via the cyclin I/ATM/ATR/p53 axis. Journal of Cellular and Molecular Medicine. 2021;25(4):1896-1910

[99] Kwan SY, Sheel A, Song CQ, Zhang XO, Jiang T, Dang H, et al. Depletion of TRRAP induces p53-independent senescence in liver cancer by Down-regulating mitotic genes. Hepatology. 2020;71(1):275-290

[100] Zhang C, Liu J, Xu D, Zhang T, Hu W, Feng Z. Gain-of-function mutant p53 in cancer progression and therapy. Journal of Molecular Cell Biology. 2020;12(9):674-687

[101] Momand J, Zambetti GP, Olson DC, George D, Levine AJ. The mdm-2 oncogene product forms a complex with the p53 protein and inhibits p53-mediated transactivation. Cell. 1992 Jun 26;69(7):1237-1245

[102] Barak Y, Juven T, Haffner R, Oren M. mdm2 expression is induced by

wild type p53 activity. The EMBO Journal. 1993;**12**(2):461-468

[103] Chen JI, Marechal VI, Levine AJ. Mapping of the p53 and mdm-2 interaction domains. Molecular and Cellular Biology. 1993;**13**(7):4107-4114

[104] Picksley SM, Lane DP. What the papers say: The p53-mdm2 autoregulatory feedback loop: A paradigm for the regulation of growth control by p53? Bioessays. 1993;**15**(10):296

[105] Grossman SR, Perez M, Kung AL, Joseph M, Mansur C, Xiao ZX, et al. p300/MDM2 complexes participate in MDM2-mediated p53 degradation. Molecular Cell. 1998;**2**(4):405-415

[106] Kussie PH, Gorina S, Marechal V, Elenbaas B, Moreau J, Levine AJ, et al. Structure of the MDM2 oncoprotein bound to the p53 tumor suppressor transactivation domain. Science. 1996;**274**(5289):948-953

[107] Brooks CL, Gu W. p53 ubiquitination: Mdm2 and beyond. Molecular Cell. 2006;**21**(3):307-315

[108] Kubbutat MH, Jones SN, Vousden KH. Regulation of p53 stability by Mdm2. Nature. 1997;**387**(6630):299-303

[109] Moll UM, Petrenko O. The MDM2-p53 interaction. Molecular Cancer Research. 2003;**1**(14):1001-1008

[110] Picksley SM, Vojtesek B, Sparks A, Lane DP. Immunochemical analysis of the interaction of p53 with MDM2;--fine mapping of the MDM2 binding site on p53 using synthetic peptides. Oncogene. 1994;**9**(9):2523-2529

[111] Xirodimas DP, Stephen CW, Lane DP. Cocompartmentalization of p53 and Mdm2 is a major determinant for Mdm2-mediated degradation of p53.

Experimental Cell Research. 2001;**270**(1):66-77

[112] Ginsberg D, Mechta F, Yaniv M, Oren M. Wild-type p53 can down-modulate the activity of various promoters. Proceedings of the National Academy of Sciences. 1991;**88**(22):9979-9983

[113] Ramqvist T, Magnusson KP, Wang Y, Szekely L, Klein G, Wiman KG. Wild-type p53 induces apoptosis in a Burkitt lymphoma (BL) line that carries mutant p53. Oncogene. 1993;**8**(6): 1495-1500

[114] Putcha GV, Harris CA, Moulder KL, Easton RM, Thompson CB, Johnson EM Jr. Intrinsic and extrinsic pathway signaling during neuronal apoptosis: Lessons from the analysis of mutant mice. The Journal of Cell Biology. 2002;**157**(3):441-453

[115] Pereira H, Silva S, Julião R, Garcia P, Perpétua F. Prognostic markers for colorectal cancer: Expression of P53 and BCL2. World Journal of Surgery. 1997;**21**(2):210-213

[116] Saelens X, Festjens N, Walle LV, Van Gurp M, Van Loo G, Vandenabeele P. Toxic proteins released from mitochondria in cell death. Oncogene. 2004;**23**(16):2861-2874

[117] Fulda S, Debatin KM. Extrinsic versus intrinsic apoptosis pathways in anticancer chemotherapy. Oncogene. 2006;**25**(34):4798-4811

[118] Shukla S, Dass J, Pujani M. p53 and bcl2 expression in malignant and premalignant lesions of uterine cervix and their correlation with human papilloma virus 16 and 18. South Asian Journal of Cancer. 2014;**3**(01):048-053

[119] Fontanini G, Boldrini L, Vignati S, Chine S, Basolo F, Silvestri V, et al. Bcl2 and p53 regulate vascular endothelial growth factor (VEGF)-mediated

angiogenesis in non-small cell lung carcinoma. European Journal of Cancer. 1998;**34**(5):718-723

[120] McCurrach ME, Connor TM, Knudson CM, Korsmeyer SJ, Lowe SW. Bax-deficiency promotes drug resistance and oncogenic transformation by attenuating p53-dependent apoptosis. Proceedings of the National Academy of Sciences. 1997;**94**(6): 2345-2349

[121] Shaw P, Bovey R, Tardy S, Sahli R, Sordat B, Costa J. Induction of apoptosis by wild-type p53 in a human colon tumor-derived cell line. Proceedings of the National Academy of Sciences. 1992;**89**(10):4495-4499

[122] Toshiyuki M, Reed JC. Tumor suppressor p53 is a direct transcriptional activator of the human bax gene. Cell. 1995;**80**(2):293-299

[123] Day CL, Smits C, Fan FC, Lee EF, Fairlie WD, Hinds MG. Structure of the BH3 domains from the p53-inducible BH3-only proteins Noxa and Puma in complex with Mcl-1. Journal of Molecular Biology. 2008;**380**(5):958-971

[124] Hemann MT, Zilfou JT, Zhao Z, Burgess DJ, Hannon GJ, Lowe SW. Suppression of tumorigenesis by the p53 target PUMA. Proceedings of the National Academy of Sciences. 2004;**101**(25):9333-9338

[125] Haupt S, Berger M, Goldberg Z, Haupt Y. Apoptosis-the p53 network. Journal of Cell Science. 2003;**116**(20): 4077-4085

[126] Song G, Chen GG, Yun JP, Lai PB. Association of p53 with bid induces cell death in response to etoposide treatment in hepatocellular carcinoma. Current Cancer Drug Targets. 2009;**9**(7): 871-880

[127] Moroni MC, Hickman ES, Denchi EL, Caprara G, Colli E,

Cecconi F, et al. Apaf-1 is a transcriptional target for E2F and p53. Nature Cell Biology. 2001;**3**(6):552-558

[128] MacLachlan TK, El-Deiry WS. Apoptotic threshold is lowered by p53 transactivation of caspase-6. Proceedings of the National Academy of Sciences. 2002;**99**(14):9492-9497

[129] de la Monte SM, Sohn YK, Wands JR. Correlates of p53-and Fas (CD95)-mediated apoptosis in Alzheimer's disease. Journal of the Neurological Sciences. 1997;**152**(1): 73-83

[130] Taketani K, Kawauchi J, Tanaka-Okamoto M, Ishizaki H, Tanaka Y, Sakai T, et al. Key role of ATF3 in p53-dependent DR5 induction upon DNA damage of human colon cancer cells. Oncogene. 2012;**31**(17): 2210-2221

[131] Puzio-Kuter AM, Castillo-Martin M, Kinkade CW, Wang X, Shen TH, Matos T, et al. Inactivation of p53 and Pten promotes invasive bladder cancer. Genes & Development. 2009;**23**(6):675-680

[132] Freeman DJ, Li AG, Wei G, Li HH, Kertesz N, Lesche R, et al. PTEN tumor suppressor regulates p53 protein levels and activity through phosphatase-dependent and-independent mechanisms. Cancer Cell. 2003;**3**(2): 117-130

[133] Rangel LP, Ferretti GD, Costa CL, Andrade SM, Carvalho RS, Costa DC, et al. p53 reactivation with induction of massive apoptosis-1 (PRIMA-1) inhibits amyloid aggregation of mutant p53 in cancer cells. Journal of Biological Chemistry. 2019;**294**(10):3670-3682

[134] Rippin TM, Bykov VJ, Freund SM, Selivanova G, Wiman KG, Fersht AR. Characterization of the p53-rescue drug CP-31398 in vitro and in living cells. Oncogene. 2002;**21**(14):2119-2129

[135] Tovar C, Graves B, Packman K, Filipovic Z, Xia BH, Tardell C, et al. MDM2 small-molecule antagonist RG7112 activates p53 signaling and regresses human tumors in preclinical cancer models. Cancer Research. 2013;73(8):2587-2597

[136] Jeay S, Gaulis S, Ferretti S, Bitter H, Ito M, Valat T, et al. A distinct p53 target gene set predicts for response to the selective p53–HDM2 inhibitor NVP-CGM097. eLife. 2015;4:e06498

[137] Wang S, Sun W, Zhao Y, McEachern D, Meaux I, Barrière C, et al. SAR405838: An optimized inhibitor of MDM2–p53 interaction that induces complete and durable tumor regression. Cancer Research. 2014;74(20):5855-5865

[138] Wodarz D. Gene therapy for killing p53-negative cancer cells: Use of replicating versus nonreplicating agents. Human Gene Therapy. 2003;14(2): 153-159

[139] Swisher SG, Roth JA, Komaki R, Gu J, Lee JJ, Hicks M, et al. Induction of p53-regulated genes and tumor regression in lung cancer patients after intratumoral delivery of adenoviral p53 (INGN 201) and radiation therapy. Clinical Cancer Research. 2003;9(1): 93-101

[140] Fujiwara T, Grimm EA, Mukhopadhyay T, Zhang WW, Owen-Schaub LB, Roth JA. Induction of chemosensitivity in human lung cancer cells in vivo by adenovirus-mediated transfer of the wild-type p53 gene. Cancer Research. 1994;54(9): 2287-2291

[141] Roth JA. Adenovirus p53 gene therapy. Expert Opinion on Biological Therapy. 2006;6(1):55-61

[142] Wong HH, Lemoine NR, Wang Y. Oncolytic viruses for cancer therapy: Overcoming the obstacles. Viruses. 2010;2(1):78-106

[143] Bressy C, Hastie E, Grdzelishvili VZ. Combining oncolytic virotherapy with p53 tumor suppressor gene therapy. Molecular Therapy-Oncolytics. 2017;5:20-40

[144] Kang SJ, Kim BM, Lee YJ, Chung HW. Titanium dioxide nanoparticles trigger p53-mediated damage response in peripheral blood lymphocytes. Environmental and Molecular Mutagenesis. 2008;49(5): 399-405

[145] Ng KW, Khoo SP, Heng BC, Setyawati MI, Tan EC, Zhao X, et al. The role of the tumor suppressor p53 pathway in the cellular DNA damage response to zinc oxide nanoparticles. Biomaterials. 2011;32(32):8218-8225

[146] Asharani PV, Xinyi N, Hande MP, Valiyaveettil S. DNA damage and p53-mediated growth arrest in human cells treated with platinum nanoparticles. Nanomedicine. 2010;5(1):51-64

[147] Malhotra P, Anwar M, Nanda N, Kochhar R, Wig JD, Vaiphei K, et al. Alterations in K-ras, APC and p53-multiple genetic pathway in colorectal cancer among Indians. Tumour Biology. 2013;34(3):1901-1911

[148] Anwar M, Nanda N, Bhatia A, Akhtar R, Mahmood S. Effect of antioxidant supplementation on digestive enzymes in radiation induced intestinal damage in rats. International Journal of Radiation Biology. 2013;89(12):1061-1070

Role of p53 as an Environmental Biomarker and the Dynamics and Energetics of p53 Binding to DNA

Presence of p53 Protein on Spermatozoa DNA: A Novel Environmental Bio-Marker and Implications for Male Fertility

Salvatore Raimondo, Mariacira Gentile, Tommaso Gentile and Luigi Montano

Abstract

Many studies suggest a direct relationship between toxic effects and an increase in the p53 protein on cellular DNA. For our studies, we used sperm DNA as an indicator of environmental toxic effects, dosing p53 quantitatively. To assess possible variations, we used semen samples from two homogeneous male groups living permanently in areas with different environmental impact. The toxic effects of the selected high environmental impact area are caused by both soil and air pollution, while the selected low environmental impact area is a nature reserve where there are no landfills, but only rural factories. As we work with reproductive cells, our interest was inevitably focused on sperm DNA damage and whether this damage could affect their fertilizing capacity. The length of telomeres and the quantification of protamines are being studied to better define the possible damage.

Keywords: p53, DFI, spermatogenesis, infertility, sperm DNA damage

1. Introduction

The combination of health and environment is now a major issue on the political agenda of many governments because of its social and cultural relevance to both individual and collective health.

The World Health Organization (WHO) has set – as one of its main priorities – the understanding of the relationship between sources of pollution and the effects on health, the development of indicators and the prevention of diseases linked to an unhealthy environment, which are a major cause of mortality and morbidity [1]. Sitography (https://apps.who.int/iris/handle/10665/204585).

Therefore, the 'eco-epidemiological' study of the determinants of health and their spatial and temporal distribution is of great interest, as these are strongly linked to social, cultural and environmental factors that mutually interact and affect the genetic heritage of individuals.

To understand which elements should be taken into account, from an epidemiological point of view, in order to assess the impact of different factors on health status is a very complex task.

The combination of environmental, territorial and epidemiological data, as well as other health, demographic, cultural and social indicators, allows us to draw up risk thresholds or possible risk scenarios for a specific population (www.epicentro.iss.it).

It's now well acknowledged that pollution plays a major role in determining an adverse health effect, and that the health condition of the population varies according to whether environmental pressure is greater or lower in an area compared to another, varying not only over countries, but also within the same country or even the same region.

Human semen is an early sensor of the environmental contamination status and therefore the first to be affected [2, 3]; Kimberley [4–6].

Chemical substances found in the environment (such as heavy metals and dioxins) in food (such as agro-pharmaceuticals or insecticides), as well as unhealthy life styles or electromagnetic pollution are the main cause of alterations of semen parameters [7, 8]. The well-known mechanisms whereby chemical and physical environmental factors, whether combined or not, interfere with reproductive function are: induction of oxidative stress, hormonal imbalance, genetic and epigenetic alterations [9, 10].

Concerning the sperm decline of the last few decades, there is much concern among researchers dealing with human reproduction. More specifically, a major meta-analysis study on data collected from 1973 to 2011 among the male population in Western countries suggest that the concentration of spermatozoa drastically decreased by more than 50%, from 99 million per milliliter to 47 million per milliliter [11] and the situation is certainly no better in some countries such as Africa, India, Brazil and China [12–15]. The decline in semen quality seems to mirror the impact that pollution and bad lifestyles have had and are still having on human health.

Usually, all forms of stress, whether endogenous or exogenous, affecting the organism lead to a response from the latter, primarily from the basic morpho-functional unit, i.e. the cell.

The cell fate decision machinery is composed of multiple complex signaling pathways, in which p53 plays a central role in coordinating the multiple cellular signaling pathways as well as determining cell fate [16, 17].

When this factor is diverted from its normal control and repair functions, the regulation of cell growth may be blocked and the cell rapidly multiplies abnormally [18, 19].

The first evidence that p53 could control cell fate was gathered from studies using a myeloid leukemia cell line [20]. The finding that p53 can lead to apoptosis was confirmed by analogous experiments in which a temperature-sensitive p53 or WT p53 was also forcibly expressed in erythroleukemia cells [21], in a colon cancer cell line [22] and in a Burkitt's lymphoma cell line [23]. The p53 protein is not essential for our survival, but its role in protecting our organism from modified cells is crucial, hence the definition of 'Guardian of the Genome', referring to its role in preserving stability by preventing mutations [24]. Since the biological role of p53 is to ensure the integrity of the genome in cells, it can stimulate repair processes and protective mechanisms, or stop cell division and stimulate induction of cell death (apoptosis) [25]. Primarily through its transcription factor function, p53 has the ability to induce cell-cycle arrest and apoptosis, both of which protect the cell and the organism from DNA damage that leads to genome instability [26]. The activity of the p53 protein is stimulated in response to DNA damage and various genotoxic insults that ultimately compromise genome integrity [27]. Following genotoxic stress, p53 decides cell fate: it may induce growth arrest, DNA repair or, in case of exposure to severe DNA damage, even induce cell death by apoptosis. The loss of p53 regulatory functions and activities are involved not only in the development of malignant diseases, but also in cardiovascular, neuro-degenerative, infectious and metabolic diseases, as well as participating in the aging process of the body. p53 is capable of binding specific reactive DNA elements, and the specificity of transcriptional activation depends on the ability of the DNA-binding domain and p53 protein to interact with the regulatory regions of certain genes. Transcriptional activation is determined by the N-terminus of p53, this contains several regions which interact with the transcriptional mechanism and recruiting factors that modify the local chromatin structure [28]. The p53 protein is mainly regulated by post-translational modifications, primarily phosphorylation, and the accumulation of p53 is the first step in response to cellular stress [29].

The N-terminus is strongly phosphorylated while the C-terminus contains phosphorylated, acetylated and sumoylated residues. N-terminal phosphorylations are important for stabilizing p53 and are crucial for acetylation of C-terminal sites, which in combination lead to the p53-mediated response to genotoxic stresses [30].

The degradation of p53 depends on the interaction between two proteins and is mediated by the proteasome. The link between N-terminal Mdm2 and C-terminal p53 leads to the degradation of p53 by Mdm2. Any alterations in the central DNA binding domain of p53 do not cancel the sensitivity of the protein to degradation mediated by Mdm2 [29, 31, 32].

In response to DNA damage, ATM kinase rapidly phosphorylates p53 at Ser15. The serine/threonine kinase Chk2 acts downstream of ATM by phosphorylating p53 at Ser20. These phosphorylated sites in the N-terminus of p53 are in proximity to the Mdm2 binding region of the protein, thus blocking the interaction with Mdm2, leading to stabilization of p53, which eludes proteosomal degradation [30]. Recent studies suggest that constitutive phosphorylation of p53 by protein kinase inhibits the regulation of sequence-specific DNA binding, oligomerisation status, nuclear import/export and ubiquitination [30]. Furthermore, constitutive phosphorylation of p53 by protein kinase C (PKC) at the C-terminal domain contributes to its degradation through the ubiquitin-proteasome pathway [33].

We studied the p53 protein by using it as a direct indicator of cellular DNA damage caused by environmental toxic factors, comparing levels in male gametes (spermatozoa) and associating them with the fertilizing capacity of the spermatozoa themselves. On average, it takes 64 days to complete spermatogenesis, but this varies from individual to individual. Spermatozoa are produced non-stop every day from puberty onwards over a lifetime [34]. This feature could be used to monitor changes in environmental impact, drug response (antioxidants) and/or lifestyle. Sperm chromatin is very compact and stable in the nucleus, unlike the structure of somatic cells. Nuclear condensation in spermatozoa is due to the replacement of about 85% of the DNA-associated lysine-rich histones with protamines, arginine-rich transition proteins [35, 36].

While histones form a ring with DNA (nucleosomes), protamines are bound to the grooves of the DNA helix, wrapping tightly around the strands of DNA (approximately 50 kb of DNA per protamine) to form tight and highly organized rings. Moreover, inter- and intramolecular disulphide bonds between cysteine-rich protamines are also responsible for the compaction and stabilization of the sperm nucleus [36, 37]. This leads to an extreme nuclear condensation and a reduction of about 10% in the size of the nucleus [35]. The BRDT protein (Bromodomain Testis specific) is the key protein that mediates chromatin compaction and can facilitate nuclear remodeling, thus ensuring the transition between the histone organization of the chromatin, or somatic, and the protamine nucleus, typical of the mature spermatozoon [38]. Specific nuclear compaction is relevant to protect the sperm genome from stressogenic insults. Indeed, both physiological and environmental stress, as well as genetic mutations and chromosomal abnormalities, can interfere with the processes of spermatogenesis. These changes can lead to an abnormal chromatin structure incompatible with the reproductive plan. The faults of genomic material found in mature spermatozoa can impair nucleus formation (defective histone and protamine substitution) and maturation, leading to DNA fragmentation (i.e. single- or double-strand breaks) and DNA integrity defects or chromosomal aneuploidy in the spermatozoa [36]. In atypical and immature spermatozoa, DNA may fragment, lose its functional integrity and thus result in functional defects in the spermatozoa. As a matter of fact, DNA fragmentation is particularly common in sub-fertile human spermatozoa [36].

p53 is one of the most investigated tumor suppressor proteins and is involved in cell cycle regulation, through its effects on transcription regulation in response to DNA damage and cell stress, resulting in DNA repair, cellular senescence, growth suppression, or apoptosis. Studies also indicate the involvement of p53 in spermatogenesis [39]. During normal spermatogenesis, p53 is expressed in the intermediate layer of the seminiferous tubules, in spermatocytes and round spermatids, suggesting that it might play a role in spermatogenesis [40, 41].

It has actually been suggested that the role of the ancestral p53 gene is to ensure the integrity of the genomic germ line and the accuracy of developmental processes [42]. The p53 protein fulfills several functions in the meiotic and premeiotic stages of spermatogenesis [43]. Possibly, p53 plays different roles in DNA repair, depending on the type of damage [44], the stage at which the cell was damaged and the possible repair pathways available [43]; in short: p53 helps the spermatozoon to deal safely with DNA damage [45].

DNA damage, resulting from normal metabolic processes in the cell, occurs at a rate of 1000 to one million molecular lesions per cell per day. Nevertheless, several causes of damage can increase this rate. Causes of alterations in sperm DNA include both extrinsic (environmental and lifestyle factors) and intrinsic causes. Apoptosis, or programmed cell death, is a natural process of cells whereby an aged or damaged cell dies without damaging its neighbors [46].

As for sperm cells, apoptosis mainly occurs to spermatogonia during spermiohistogenesis, a significant factor in blocking the complete development of a damaged cell. Apoptosis also occurs in mature spermatozoa when they manifest alterations that could be passed on to their offspring or that hinder the normal functions of the cell itself [47].

Many studies have been carried out over the years to assess the harmful effects of environmental factors on sperm DNA. The first studies were carried out on the effects of cigarette smoking and new techniques were developed to highlight the damage [48]. When comparing the DNA fragmentation index of spermatozoa from smoking and non-smoking patients, researchers were able to determine that the DNA damage detected in smokers was greater [49]. DNA breaks can be caused by the presence of carcinogens and mutagens in cigarette smoke [50]. Harmful substances, including alkaloids, nitrosamines, nicotine, cotinine and hydroxycotinine are found in cigarettes and produce free radicals [51]. Kunzle et al. [52], an association between cigarette smoking and sperm quality was found among extrinsic causes, i.e. due to environmental factors. Rodgman and Perfetti [53] and Alchinbayev et al. [54] highlight mutagenic properties of cigarette constituents and altered sperm quality.

Oxidative stress (OS) is the focus of in-depth studies, due to the potential harmful effects of high levels of reactive oxygen species (ROS)[55]. An increase in leukocytes is supposed to determine an increase in ROS production in semen but the process is still not very clear [56].

Environmental toxic effects damage sperm nuclear and mitochondrial DNA. The assessment of damage related to non-functional spermatozoa is extremely significant for male fertility [57]. Sperm DNA damage reaches higher levels in infertile men than in fertile men and, as a matter of fact, more and more studies prove that sperm DNA damage negatively affects reproductive outcomes [58]. These damages may not only impair fertility, but also increase the transmission of genetic diseases during ART procedures [59]. Spermatozoa produce small amounts of ROS and these play a significant role in many sperm physiological processes, such as capacitation, hyperactivation and sperm-oocyte fusion [60, 61]. However, ROS must be inactivated continuously to keep only a small amount necessary to preserve normal cell function. Overproduction of ROS in semen can result in sperm DNA damage. An overproduction of ROS in semen can result in sperm DNA.

During their maturation process, spermatozoa extrude their cytoplasm, the main source of antioxidants. Once this process is slowed down, the residual cytoplasm forms a cytoplasmic droplet in the sperm mid region. These spermatozoa carrying cytoplasmic droplets are immature and functionally defective [62]. The residual cytoplasm contains a high concentration of certain cytoplasmic enzymes (G6PDH = Glucose-6-Phosphate DeHydrogenase, SOD = SuperOxide Dismutase), which are also a source of ROS [62]. The lack of cytoplasm leads to a decrease in antioxidant defense. This process is the link between poor sperm quality and high levels of ROS [56, 63].

Human ejaculate consists of different cell types: mature and immature spermatozoa, round cells from different stages of the spermatogenic process, leukocytes and epithelial cells. Peroxidase-positive leukocytes and abnormal spermatozoa continuously produce free radicals. Spermatozoa are extremely sensitive to damage caused by excessive ROS because their cytoplasmic membranes contain large amounts of polyunsaturated fatty acids (PUFAs), which intensify lipid peroxidation by ROS, resulting in a loss of membrane integrity [55, 64, 65]. There is a strong positive correlation between immature spermatozoa and ROS production, which in turn is negatively connected to semen quality [66]. Moreover, the concentration of mature spermatozoa with damaged DNA was found to increase along with immature spermatozoa in the human ejaculate [47].

Over the last few decades, scientific evidence of the harmful effects on spermatogenesis of occupational exposure chemicals known as endocrine disruptors (EDCs) on the reproductive system has been progressively accumulating [67, 68]. Environmental pollution is one of the main sources of ROS production and has been involved in the pathogenesis of poor semen quality [69]. A study carried out on workers at motorway toll booths, who are constantly exposed to environmental pollutants, correlated blood methaemoglobin and lead levels in semen were inversely correlated, compared to local male inhabitants not exposed to heavy traffic pollution levels. These results suggest that nitrogen oxide and lead, both found in the composition of car exhaust, negatively affect semen quality [70]. Furthermore, increased industrialization has led to a high deposition of highly toxic heavy metals in the atmosphere. Paternal exposure to heavy metals such as lead, arsenic and mercury is associated with a decrease in semen parameters, resulting in a reduced fertility capacity [71, 72].

Global pollution was negatively associated with sperm count in a group of Californian sperm donors. This study shows a significantly negative relationship between sperm concentration and ozone levels measured 0–9, 10–14 and 70–90 days prior to semen collection. Since ozone appears not to be involved in oxygen transport mechanisms, the mechanism of action remains to be clarified, although the observed effect reinforces the evidence on the relationship between spermatogenesis and traffic-related pollution [73].

As for pesticides to which the population is exposed or has been exposed in the past, the available results of specific studies on their effects on spermatogenesis are still inconsistent. This also applies to the well-known DDT, which is now banned in Western countries: the effect of this pesticide on spermatogenesis is low [74].

Reproductive capacity, on the other hand, does not seem to be adversely affected other than marginally [75, 76].

Similar considerations apply to other persistent contaminants in the environment. Contrary to this general consideration, an American study reported a highly significant association between urinary levels of the metabolites of three pesticides and a reduced number of spermatozoa in the ejaculate. However, this study also found a decrease in the number of spermatozoa, albeit insignificant [77].

However, Marty et al. [78] found no qualitative differences in the incidence of abnormalities in spermatozoa form and number related to p53 concentration, in contrast to the data reported by Yin et al. [79]. The latter reported that the p53 protein controls germ cell quality by inducing spontaneous apoptosis, failure to do so results in the accumulation of defective cells, which increases the concentration of abnormal spermatozoa and subsequently compromises male fertility. These data are supported by more recent studies reporting a negative correlation with nemaspermic motility [80]. Sperm vitality correlates strongly with the DNA fragmentation index [81] and oxidative stress, caused by harmful environmental exposure, is believed to have a significant role in the development and progression of diseases [82].

The function of p53 to govern the fate of cellular life, when it is damaged, is now well known. p53 monitoring is useful for assessing the effects of pollutants on DNA. Considering the changes of p53 in relation to the degree of the DNA damage, quantitative measurement of the p53 protein on sperm DNA was performed to evaluate:

a. possible negative effects of pollutants on male fertility in subjects living in high environmental impact area;

b. possible sperm DNA damage following manipulation of spermatozoa during the separation procedures for ART techniques, evaluating the quality of the embryos too.

For this aim, the method proposed by Raimondo et al. [83] consists of 3 steps:

1. separation of spermatozoa from seminal fluid using a forensic method [84].

2. isolation of nuclear DNA from spermatozoa.

3. quantitative evaluation of the p53 protein by ELISA.

The concentration of the spermatozoa is reported in Mil/ML, 100 micronliters of seminal fluid are used for the p53 protein assay, therefore the p53 protein concentration is correlated to 1/10 of the sperm count per ML. The correlation existing between p53 concentration and number of spermatozoa per ML, allows us to report the p53 values in "p53 ng/million spermatozoa" [83].

$$\text{Corrected p53} = \frac{\text{Value of p53 ng} / 100 \text{ micronliters}}{1/10 \text{ of the spermatic count} / ml}$$

p53 protein values are expressed in ng/million spermatozoa.

2. p53 concentration on sperm DNA and environmental impact

The ancestral p53 gene is involved in ensuring the integrity of the genomic germline and the replication of developmental processes. The p53 protein is highly expressed in testicles, spermatogonia and primary spermatocytes during pachytene or pre-leptotene, when chromosome pairing, recombination and DNA repair occur. The expression of p53 at these stages of spermatogenesis suggests that it plays a role in meiosis. Apoptosis is a critical process for the integrity of germ cell DNA and in regulating their quantity.

If p53 concentrations are not adequate, this would lead to aberrant spermatogenesis or sperm containing damaged DNA. Failure to control p53 leads to the accumulation of defective cells, which increases the concentration of abnormal spermatozoa [85] and subsequently impair male fertility. These data are supported by more recent studies reporting a negative correlation with nemaspermic motility [81, 86]; additionally, sperm vitality is strongly correlated with DNA fragmentation index [87, 88].

We carried out an observational study on 117 male subjects, aged 18–38 years (28.02 + 4.99), permanently living in low and high environmental impact areas from July 2015 to June 2020.

Our purpose is to assess the concentration of the p53 protein on spermatozoa DNA using an immunoenzymatic assay (ELISA) as a marker of possible damage. The whole group consisted of 117 males divided as follows: 49 of them permanently living in low environmental impact areas (southern area of Salerno; Campania, Italy), aged 18–38 (28.04 + 4.84) years identified as Group A; 68 of them permanently living in high environmental impact areas (northern area of Naples 'terra dei fuochi'; Campania, Italy), aged 18–37 (28.01 + 5.13) years identified as Group B. The observation lasted 60 months, among the requirements: homogeneous behavior and lifestyle, no habitual smokers, no alcohol abusers and except for some of them who has used cannabis in the past (whose suspension is reported from 6 to 36 months before the collection of semen), they do not perform activities considered to be an environmental occupational risk and did not suffer from pathological varicocele at preliminary examination with Color Doppler [50, 55, 59, 89, 90].

The examination of the human semen was evaluated using the standardized analysis criteria according to the WHO Laboratory Manual for the examination and processing of human semen, fifth edition – 2010. In Group A, the ejaculate volume ranged from 1.1 to 4.9 mL, and the seminal evaluations were as follows: 24 samples (48.9%) normospermic; 14 (28.6%) mild oligospermic; 7 (14.3%) medium oligospermic; 4 (8.2%) severe oligospermic. In group B, the ejaculate volume varies from 0.6 to 7.1 ml, the seminal evaluations were as follows: 13 (19.1%) normospermic; 20 (29.4%) mild oligospermic; 27 (39.7%) medium oligospermic; 8 (11.8%) severe oligospermic. The Makler Counting Chamber (Sef-Medical Instrumens ltd.) was used to evaluate the nemasperm concentration expressed per ml, the number of spermatozoa as well as the study of the non-nemaspermic or immature nemaspermic cellular component (leukocytes, red blood cells, germ line cells) (**Table 1**) [49, 53, 91].

Sample processing procedures were carried out 30 minutes after ejaculation. Samples were divided into two aliquots, one of which was processed immediately for the p53 ELISA assay and the other frozen at −20° for later examination. A quantitative assessment of p53 corrected according to the number of spermatozoa was performed on all samples and values are expressed in ng/MLN spermatozoa. The method employed was that suggested by Raimondo et al. [83].

Data suggest that there are significant differences in seminal parameters from groups A and B.

These variations are probably due to the effects of environmental factors on the organism, and on semen in particular (**Figure 1**). This finding is further supported by the fact that the examined groups are homogeneous, as previously reported.

	Low environmental impact Group A	High environmental impact Group B
MLN spermatozoa/mL	41.26 ± 14.6	27.12 ± 9.8
Motility type (a)	33.7 ± 11.5	28.1 ± 9.6
Morfology	15.6 ± 2.8	13.8 ± 3.8
Vitality	61.2 ± 6.3	57.4 ± 8.1

Table 1.
Description of seminal parameters of the two groups.

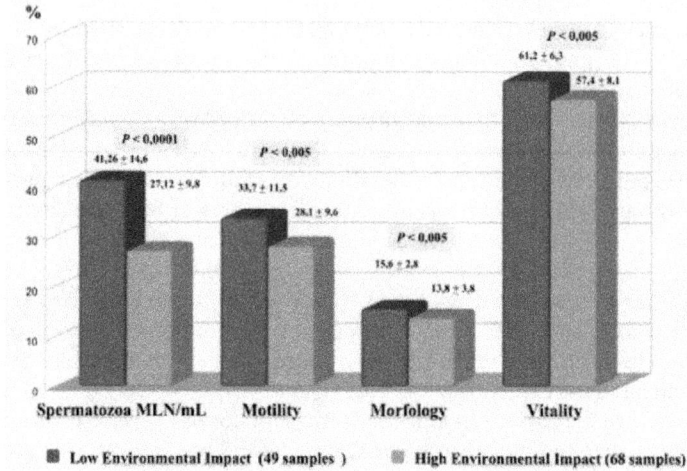

Figure 1.
Significant variations in the main seminal parameters in groups A and B.

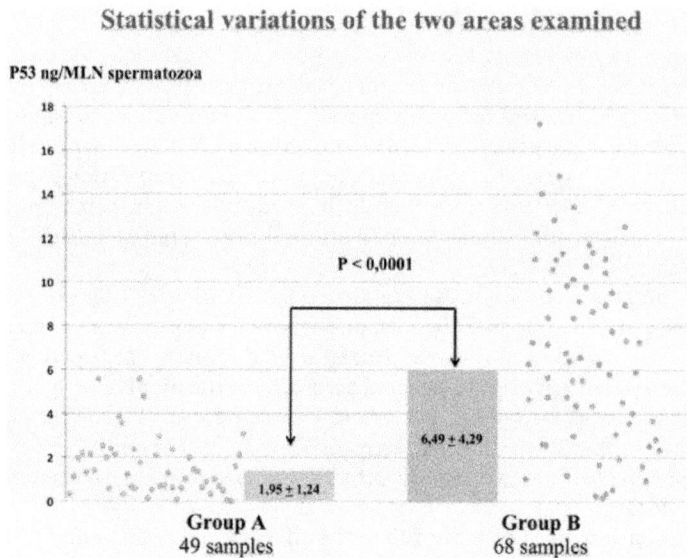

Figure 2.
Statistical changes in p53 values in the two areas under examination.

In order to assess the possible damage to the spermatozoa DNA, we used the quantitative analysis of the p53 protein and results show a significant variation ($p < 0.0001$) between the two groups: group A; p53= 1.95+1.24: group B; p53= 6.49+4.29 (**Figure 2**).

These data highlight that environmental factors are strongly associated with seminal parameters alteration and with sperm DNA damage in subjects living in high environmental impact areas and, inevitably, these alterations may interfere with the reproductive plans of couples living in these areas.

3. p53 concentration on sperm DNA and male fertility

Spermatogenesis is male gametogenesis, i.e. the maturation process of male germ cells that takes place in testicles under the stimulus of the hormones FSH and testosterone when the individual has reached puberty. Although it's the equivalent of oogenesis in women, it differs from the latter mainly in terms of timing: sperm production begins at puberty and lasts a lifetime, oogenesis begins before birth and then stops and resumes when the woman reaches sexual maturity, ending at menopause. Spermatogenesis is not to be confused with spermiogenesis, which is the third and final stage of spermatogenesis, during which the final differentiation takes place, leading to the development of mature spermatozoa [92, 93].

At the end of spermatogenesis, only 15–20% of spermatozoa are normal, the residual being functionally or morphologically abnormal spermatozoa.

Spermatogenesis takes place inside the testes and more precisely in the seminiferous tubules, which are blind-ending tubules that converge in the recti seminiferous tubules. The tubules recti then converge to form the *rete testis*, from which 15–20 efferent ducts drain into the epididymis and then continues into the vas deferens. The wall of these seminiferous tubules consists of supporting cells, called Sertoli cells, and various germ cells that make up the various stages of spermatogenesis. The duration of spermatogenesis can take 70 to 90 days and begins with the division of undifferentiated cells located near the basal lamina of the seminiferous tubule (spermatogonia). These cells undergo mitosis and meiosis, resulting in the production of mature cells (spermatozoa) which detach from the most luminal part of the tubular wall. Germ cells then undergo a process that brings them from the most marginal regions of the wall towards the most apical regions, until they are released into the lumen of the tubule. The cells involved in spermatogenesis are divided into two large groups: germ cells, consisting of spermatozoa and their precursors, and non-germ cells, cells that are not precursors and never become gametes, but have trophic and regulatory functions [92, 93]. The primordial germ cells settling in male gonads form hollow structures called seminiferous tubules, whose wall consists of somatic cells called Sertoli cells. Outside the seminiferous tubule, within the connective tissue that surrounds it, there are the Leydig cells, responsible for the production of testosterone. In this situation, the germ cells, represented by the A1 type spermatogonia, which have already undergone a cellular multiplication during organogenesis, remain dormant until sexual maturity. The Sertoli cells are tightly connected to each other in the basal compartment by occluding junctions that together form the blood-testicular barrier. This barrier means that the seminiferous tubule is structured into two compartments: the basal compartment (housing the spermatogonia and the leptotene spermatocytes) and the adluminal compartment (housing the more mature spermatocytes, spermatids and spermatozoa) [94, 95].

The blood-testicular barrier has several functions: it ensures the preservation of distinct microenvironments between the two compartments so as to help meiosis and spermiohistogenesis in the adluminal compartment and prevent possible immunological responses following exposure to germ antigens or the transit of macromolecules from the adluminal compartment into the bloodstream.

The spermatogenesis is a complex process in which differentiation and mitosis of a group of starting stem cells take place. The germ cell is called a spermatogonium and divides by mitosis into two cells. The first is a differentiated spermatocyte while the second maintains the features of spermatogonium, to ensure the turnover of the germ cell base [96]. The primary spermatocyte is different from the spermatogonium and takes part in the meiosis process. During the first stage, the primary spermatocyte (a diploid) divides into two secondary spermatocytes (haploids) containing half the genetic patrimony of the primary spermatocyte. The newly formed secondary spermatocyte is still in the meiosis stage and with the second reduction, not reducing its genetic patrimony, it divides into two spermatids. Each spermatid is then 'refined' inside the gonad because it is not yet capable of undergoing fertilization. The 'refining' is to be understood as a variable length process, aimed at creating and reinforcing the structure of the future spermatozoon, which requires particular elements that are not present in the spermatids in order to fulfill its task [97]. At the final stage, the spermatozoon has a typical structure: mature spermatozoon [98].

The p53 protein was found to have several functions in the meiotic and pre-meiotic stages of spermatogenesis [99]. Possibly, p53 plays different roles in DNA repair, depending on the type of damage, or on the stage at which the cell was damaged, and on the possible repair pathways available [42]. The p53 protein helps sperm to deal safely with DNA damage [100]. A study by Lane shows that p53 plays a role in spermatogenesis: as a matter of fact, mRNA and p53 protein are highly expressed during mouse and rat spermatogenesis and we deal with primary pre-myiotic spermatocytes at the zygotene-pachytene stages, before the beginning of meiotic division [101]. In addition, p53-knockout mice and mice with reduced levels of p53 show germ cell degeneration during the meiotic prophase, which occurs with the appearance of multinucleated giant cells [102]. p53 knockout mice show a higher incidence of testicular cancer, suggesting that p53 plays a role in the prevention of carcinogenesis during the mitotic stages of spermatogenesis [102–104]. p53 is also capable of mediating stress-induced apoptosis of spermatogonia after DNA damage and after overheating of testicular tissue [105]. The role of p53 in the stress response of spermatogonia is also supported by the extreme reactivity to chemo- and radio-therapy of testicular cancer cells expressing wild-type p53 [106–108]. This has been proven to be the result of the activation of 'normally latent' wild-type p53, which in turn induces a wide apoptotic response [109]. Several studies report the role of the p53 protein in the pre-meiotic and meiotic stages of spermatogenesis [110]. Recently, it has been shown that the accuracy of meiotic crossing over at different genomic locus does not cause severe difficulties in p53 knockout mice [111], moreover, the DNA damage in spermatogonia that induces apoptosis is p53 dependent, the meiotic quality control of chromosomes at meiotic metaphase I is p53 independent. On the other hand, it has been observed that knockout mice for both p53 and ATM genes proceed to later stages of meiosis than those knockout mice with only the ATM gene. Yin et al. [79] reported that p53 mice had impaired apoptosis especially in the tetraploid DNA state. These results suggest that DNA damage at the meiotic stage is p53 dependent.

The proper presence of the p53 protein in spermatogenesis ensures both the quality and the right amount of mature spermatozoa necessary for successful conception. In this observational study, we evaluate the possible correlation between p53 concentration on human sperm DNA and male fertility potential.

Our report is based on an observational study involving 169 males over a period from March 2014 to February 2019. The whole group consists of 208 male partners aged 26–38 years with ejaculate volume from 0.6 to 5.8 ml and heterogeneous

seminal evaluation: 86/208 (41.3%) normospermic; 19/208 (9.1%) mild oligospermic; 51/208 (24.5%) moderate to oligospermic; 52/208 (25.1%) with severe oligospermic. The 'control A' group includes 39 male partners considered 'fertile' because
we performed the p53 test on their sperm DNA 28 ± 3.5 days after the positive
pregnancy test results of their partners (betaHCG> 400 m U/mL). Group B, divided
into B1, B2 and B3, includes 169 male partners and was observed over a period of
60 months. These partners do not report previous conceptions, do not smoke, do
not abuse alcohol, do not use drugs and do not suffer from pathological varicoceles
examined with Color Doppler. The whole group includes married and stable cohab-
iting partners over a period of 27–39 months, reporting frequent unprotected sex.
The p53 values were corrected with respect to spermatozoa concentration, therefore,
expressed in ng/million spermatozoa, hence called 'corrected' p53 values.

3.1 Results

Group A (39 males) shows 'corrected' p53 values ranging between 0.35 and
3.20 ng/million spermatozoa and group B (169 males) shows values ranging
between 0.68 and 14.53. In group B over the observation period we recorded 21
pregnancies with initial 'corrected' p53 values ranging from a minimum of 0.84 to a
maximum of 3.29. In subgroup B1, 8 spontaneous pregnancies were obtained from
male partners with a 'corrected' p53 concentration ranging between 0.84 and 1.34.
In subgroup B2, 13 pregnancies were obtained from male partners with a 'corrected'
p53 concentration ranging between 1.66 and 3.29. In subgroup B3 (121 males) there
were neither pregnancies nor miscarriages and the 'corrected' p53 values ranged
between 3.58 and 14.53.

3.2 Conclusion

The results show that participants in group A had 'corrected' p53 values between
0.35 and 3.20 and are considered 'fertile', although 3 miscarriages occurred over the

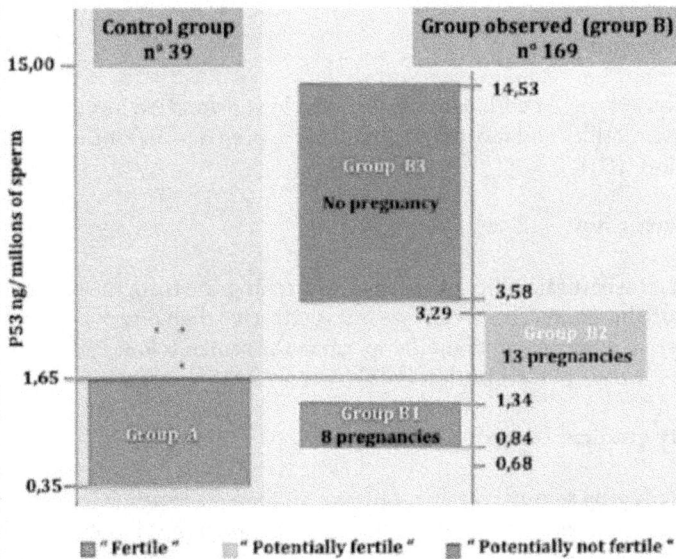

Figure 3.
Group A and B with relative "corrected" p53 concentrations. Spontaneous pregnancies with relative p53 values are reported.

observation period, 36 out of 39 males (92.3%) had a p53 concentration of less than 1.65. Participants in group B1 had a 'corrected' p53 concentration ranging between 0.84 and 1.34, with 8 pregnancies. In group B2 the 'corrected' p53 concentration ranged between 1.65 and 3.29 and 13 pregnancies were observed, so this group can be considered 'potentially fertile'. In group B3 (121 males) with 'corrected' p53 values ranging between 3.58 and 14.53, neither pregnancies nor miscarriages were observed, so it was considered 'potentially infertile' (**Figure 3**).

4. p53 concentration on sperm DNA and sperm separation techniques (ART)

Many factors damage sperm DNA. Considering an increase in the use of assisted reproduction techniques, we would like to assess whether separation techniques can be counted among the probable causes of sperm DNA damage. Spermatozoa can be isolated for several reasons: for medically assisted procreation (MAP) or diagnostic tests [112]. In MAP, the techniques for separating spermatozoa are different and all of them aim to improve the pregnancy rate (PR). The need to select/separate spermatozoa is necessary in cases of infertility due to reduced seminal parameters or to avoid the transmission of sex chromosome diseases. The ideal technique for separating spermatozoa should be easy, fast and affordable, should allow the highest number of motile spermatozoa to be isolated, should not damage or physiologically alter the spermatozoa, should eliminate non-viable spermatozoa, leukocytes and bacteria and should allow selection in the event of hyperspermia (increased ejaculate volume). Currently, no technique meets all these requirements, so the choice of sperm preparation technique is dictated solely by the embryologist's ability and the quality of the semen [113, 114].

The three spermatozoa separation techniques considered in our work are some of those reported in the 5th edition of the WHO Laboratory Manual for the examination of human semen and are also the most frequently used in MAP (Medically Assisted Procreation) centres:

4.1 Direct swim-up

This requires semen with parameters at the lower standard limits for sperm number and motility and is often used for sperm preparation for intrauterine insemination (IUI).

4.2 Pellet swim-up

Exploits the natural ability of spermatozoa to migrate from the seminal plasma to the culture medium. This technique is less effective than direct swim-up, but is useful when the percentage of motile sperm in the semen is low. Pellet swim-up is often used for in vitro fertilization (IVF).

4.3 Density gradient centrifugation

By centrifuging seminal plasma, cells are separated according to their density. Moreover, motile spermatozoa actively migrate through the gradient forming a pellet at the bottom of the test tube. Usually, a two-layer discontinuous gradient with 40% density in the upper layer and 80% density in the lower layer is used. This technique is mostly used for sperm-deficient ejaculates and for ICSI (IntraCytoplasmic Sperm Injection).

In order to assess whether separation techniques can lead to spermatozoa DNA damage, we analyzed samples before and after selection procedures (DGC, pellet swim-up and direct swim-up), comparing data with pre-treatment values (control). To this end, we used an innovative technique able to quantify spermatozoa DNA damage. The reference technique is the one proposed by Raimondo et al. [83], the quantitative assessment of p53 protein on spermatozoa DNA corrected for sperm concentration. We used an Enzyme-Linked Immunoassay (ELISA), a technique that best meets laboratory requirements for accuracy, reliability and repeatability.

4.4 Population enrolled

For this study, we enrolled 63 males in the period from January 2016 to December 2019, aged 24–31 years, the volume of their ejaculates varies from 2.6 to 4.6 mL and have various patterns of dispermia. The sperm evaluations of the subjects were carried out by examining their semen using the standardized analysis criteria according to the WHO laboratory manual for the examination and processing of human semen, 5th edition, 2010.

The Makler Counting Chamber (Sef-Medical Instrumens Ltd.) was used for the assessment of nemaspermic concentration, expressed per mL, as well as for the study of the non-nemaspermic cellular component (leukocytes, red blood cells, germ line cells) [49, 53, 91].

Enrolled subjects do not suffer from chronic diseases, have not used drugs and medications during the 6 months prior to semen collection, are not exposed to environmental stress at work [115–117], did not suffer from pathological varicocele at preliminary examination with Color Doppler [118–120].

Semen samples were processed when liquefied within 30 to 45 minutes after ejaculation.

The samples were then aliquoted into four 0.5 mL aliquots and immediately processed.

The four aliquots were processed as follows:

4.5 Group (a): control

Control samples were quantitatively assessed for p53 protein at both 0 and 60 minutes. During this period of time, semen is not treated, incubated at 37°C at 5% CO2, in a 15 mL Falcon tube.

4.6 Group (b): direct swim-up

An aliquot of semen is placed under the 300 μL layer of culture medium (Quinn's, SAGE, USA). The test tube is placed at a 45° angle to increase the contact surface of semen and medium for 30 minutes at 37° C in a 5% CO_2 incubator. The supernatant fraction is removed and sent for further assessment [121, 122].

4.7 Group (c): pellet swim-up

A 0.5 mL aliquot of the whole sperm is gently mixed with 1.0 mL of sperm culture medium supplemented with 0.1% human serum albumin (Sigma-Aldrich. St. Louis, Catalog – A1653), heated to 37° C, in a 15 mL Falcon tube and centrifuged at 200 g for 8 minutes. The supernatant is discarded and the precipitate (pellet) is mixed with 1.0 mL of culture medium and centrifuged at 100 g for 45 minutes, the supernatant discarded, 300 micronL of culture medium is gently layered onto the final pellet. The test tube is placed at a 45° angle for 30 minutes at 37° C in a 5% CO2 incubator. The supernatant fraction is removed and sent for further assessment [123, 124].

4.8 Group (d): density gradient centrifugation (DGC)

80/40 gradients (Pureception, SAGE, USA) were placed in 15 mL Falcon tubes, followed by layering of 0.5 mL of whole ejaculate and then centrifuged at 200 g for 20 minutes. The gradient is removed and the pellet is washed twice (200 g x 5 minutes) with 1.0 mL of culture medium. The final pellet is layered on the surface with 300 µL of culture medium and placed at 37°C in a 5% CO2 incubator for 30 minutes. The supernatant fraction is removed and sent for further assessment [125, 126].

All samples are subjected to a quantitative assay of p53 protein corrected in relation to the number of spermatozoa.

Separation of spermatozoa is an important step in ART techniques. Our data show that the Density Gradient Centrifugation (Group d) and Direct Swim-up (Group b) techniques provide superior quality in terms of motility, vitality and apoptosis indices compared to other conventional techniques. In Group (b), apoptosis is superimposable to that of Group (d), while motility and vitality are slightly lower. Group (c) has lower parameters than the other techniques. With regard to the assessment of the p53 protein, the results are in contrast with those of apoptosis: in Group (d), the values are significantly higher than the other techniques (**Table 2**).

The mean percentage of apoptotic spermatozoa in the processed samples was evaluated by the AO test [48] and samples processed by pellet Swim-up (Group c) were found to be significantly higher than those processed by density gradient (Group d) and direct Swim-up (Group b). The lower percentage of apoptotic spermatozoa found in Group (b) and Group (d) suggest that these techniques result in a supernatant with fewer spermatozoa with fragmented DNA. The use of apoptotic spermatozoa during ART may be one of the causes of failure of MAP cycles. The negative association between sperm apoptosis and fertilization rate has been documented with several studies [127, 128]. The selection of non-apoptotic spermatozoa should be one of the most important requirements for achieving optimal conception rates in ARTs [128]; it is beyond doubt that to achieve this important parameter, it is necessary to choose a separation technique that comes closest to natural selection.

This work suggests that the spermatozoa preparation techniques commonly used for assisted reproduction techniques result in different levels of apoptosis and spermatozoa DNA damage, which can be assessed by quantifying the p53 protein isolated from spermatozoa DNA. In the future, we plan to use p53 quantization to assess the damage already existing in spermatozoa DNA of potential

	P53 ng/Mln spermatozoa		
	Before	After	P value
Control	2,72 ± 0,0	3,17 ± 2,1	NS
Direct swum-up	2,72 ± 0,3	3,18 ± 2,9	NS
Pellet swim-up	2,72 ± 0,2	4,02 ± 3,2	P<0,001
Density gradient centrifugation	2,72 ± 0,3	7,87 ± 3,9	P< 0,0001

Table 2.
Variation in p53 protein concentrations, before and after the separation technique used, including statistical changes.

patients wishing to undergo assisted reproduction techniques, so as to prevent the final result from being further compromised. In case the p53 concentration in the untreated samples is already high, a possible therapy could be evaluated for such patients to improve the starting conditions of spermatozoa thus achieving a better result [5, 6]. This work fits well into a scenario of spermatozoa quality assessment and the importance of having an objective and repetitive data prior to conception both in vivo and in vitro [129].

5. p53, embryo quality and pregnancy rate

The p53 protein is thought to play an important role in oocyte maturation, blastocyst development and embryo implantation in human reproduction [130].

p53 protein expression is low in zygotes and at the cleavage stage, but then increases around the blastocyst stage. Blastocysts from in vivo fertilization have low concentrations of p53 protein, while p53 expression is higher in embryos produced by in vitro fertilization. These findings suggest that embryo culture leads to accumulation of p53 protein transcription activity in blastocysts and may be one of the reasons for the delayed growth of embryos. Human embryos generated by intracytoplasmic sperm injection (ICSI) have a high nuclear p53 expression, associated with delayed embryo development [131]. From these considerations, a more complex role for the p53 protein emerges, which is different from just controlling the integrity of sperm DNA; it is assumed to control the timing and mode of embryo development [132].

The p53 protein plays an important role in the cell and is normally found in all cell types in the human body. It plays a central role in an extensive control network of proteins that enable the 'healthy' condition of a cell and of cellular DNA. The p53 protein is the 'director' of a well-orchestrated cell damage detection and control system. When damage occurs, the activity of the p53 protein is crucial in deciding whether to repair it or induce cell death. The death of a cell that has suffered severe DNA damage is vital for the organism because it prevents the reproduction of cells with dangerous and harmful mutations and, in the event of conception, prevents abnormal embryonic development [133].

Its increase is proportional to cellular damage, so its quantitative assessment indicates DNA damage. Also interesting is its role in controlling and regulating the meiosis process of spermatogenesis and its function in monitoring embryonic development.

The idea that the p53 protein performs multiple tasks in systemic cellular control and development and in the control of human reproductive project is gaining momentum. Our work fits well with the knowledge of the presence of the p53 protein in differentiating male fertility.

For our study, we enrolled 117 partners of couples who had undergone medically-assisted procreation (MAP) for conception.

The seminal parameters were assessed according to the criteria of the WHO 2010 manual, shown in **Table 3**.

Participants were assessed for the concentration of the p53 protein on sperm DNA, first proceeding to a DNA extraction using a forensic method and then to a quantization of the p53 protein using ELISA-immunoassay technique, with another calculation of the results, and expressed in ng/MLN spermatozoa [83].

The embryologist chose the MAP technique to be performed, based on the seminal parameters obtained after the capacitation procedure and, in order to ensure consistency in comparison, it was the same for all samples (Percoll gradients): 90 couples (76.9%) using the IVF technique and 27 couples (23.1%) using the ICSI technique.

IVF stands for In Vitro Fertilization with Embryo Transfer and is the first arti-
ficial insemination technique used. IVF is recommended for couples with proven
fertility problems: for women, especially those suffering from tubal pathologies
(obstruction of the fallopian tubes), and for men when there are minor problems
with the semen. This technique can be used mainly in patients who have already
conceived naturally, because the ability of the sperm to spontaneously penetrate
the egg cell is more certain. With IVF (or in vitro fertilization), conception takes
place outside the woman's body: the sperm spontaneously penetrate the egg cell, but
everything takes place in a test tube.

ICSI stands for IntraCytoplasmic Sperm Injection and is used in patients of
advanced maternal age (>36 years), in cases where oocyte production is low or,

Participants	MLN spermatozoa/mL	Type a motility %	Morfology % according to Kruger	Vitality %
117	20,14 ↔ 48,31	18,5 ↔ 51,6	8,5 ↔ 17,8	48,8 ↔ 76,5

Table 3.
Seminal parameters of the participants.

	1.145 < p53 > 2.45 ng/Mln spermatozoa	3.20 < p53 > 7.75 ng/Mln spermatozoa
Participants	51	66
No. MII oocytes	380	257
Embryos	248 (65.4%)	104 (40.5%)
Pregnancies	28 (PR=54.9%)	13 (PR=19.7%)

Table 4.
*Number of participants, number of total (MII) oocytes, number of embryos that reached the 6–8 cell stage,
pregnancies achieved (PR) for two groups of p53 values.*

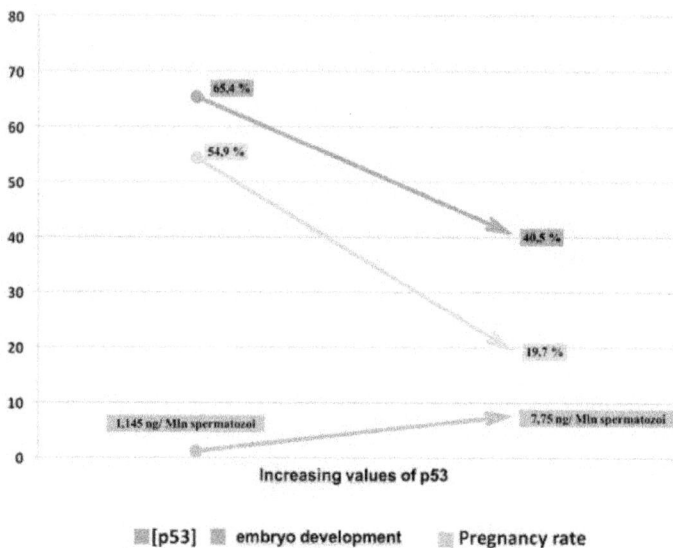

Figure 4.
Interrelation between p53 concentration, embryonic development and PR.

in the case of men, if there is severe seminal damage, such as the total absence of spermatozoa in the ejaculate fluid and it is necessary to proceed with the aspiration of sperm directly from the testicle.

The initial phase of ICSI is the same as that of IVF, starting with hormonal stimulation and then moving on to oocyte aspiration. The difference is that in ICSI a spermatozoon is selected by the biologist and injected into the cytoplasm of an oocyte using a micro needle to 'force' fertilization. This operation is repeated for all the oocytes to be inseminated. The following stages are exactly the same as IVF.

On the third day of embryo development, the number of embryos that reached the stage of 6–8 was assessed. Pick-up report (IVF + ICSI), fertilization and Pregnancy rate (PR) are shown in **Table 4**.

The results obtained support the theory that a high concentration of the p53 protein in spermatozoa DNA is associated with a low percentage of embryos able to reach the 6–8 cell stage on day three and a lower pregnancy rate (**Figure 4**).

Our work fits well with prediction models and the importance of having objective and repetitive data prior to conception, both in vivo and in vitro [134].

6. Conclusions

Cytochemistry, fluorescence and electrophoresis techniques have so far been used to assess DNA damage. For our studies, we employed an innovative method called 'quantitative proteomics', an analytical chemical technique for determining the amount of protein in a given sample. The methods for identifying proteins are identical to those used in general proteomics, but include quantification as an additional dimension. We used p53, a protein already known as the 'guardian of the genome', to assess the effect of environmental and/or dietary toxic factors on human bodies through DNA damage. From our studies, we have identified the spermatozoon as a sentinel cell of environmental impact, as its DNA damage is strongly correlated with pollution. Inevitably, the evolution of these preliminary

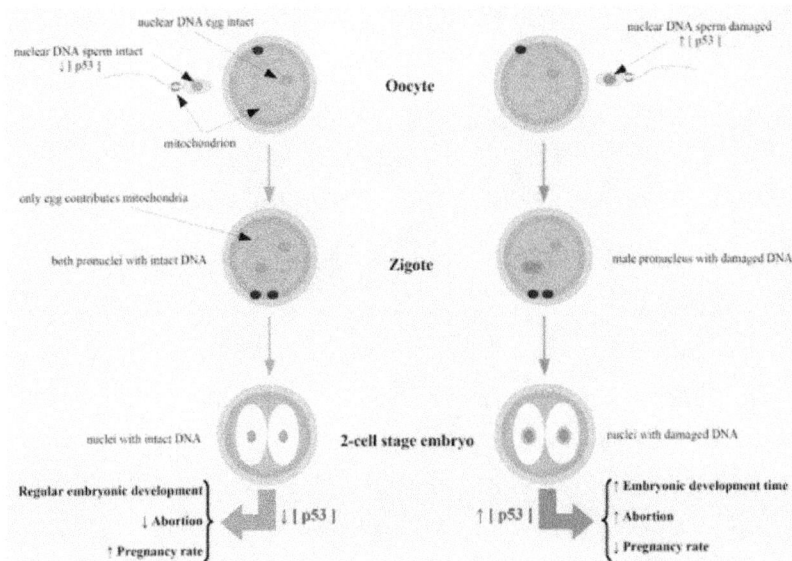

Figure 5.
Schematic representation of the effects of the different concentrations of the p53 protein on human reproduction.

studies turned to understanding whether DNA damage could influence the fertilizing capacity of males. We think that given our results, this protein can be used as an indicator of environmental impact, and given the renewal characteristics of spermatogenesis, it can also be used as a prevention and follow-up index for environmental remediation. A more extensive use would be to understand whether sperm DNA is compatible with the couple's optimal reproductive project both in vivo and in vivo (**Figure 5**).

Author details

Salvatore Raimondo[1,2]*, Mariacira Gentile[1,2], Tommaso Gentile[1,2] and Luigi Montano[2,3]

1 Laboratory "Gentile s.a.s." Research Department, Gragnano, Naples, Italy

2 Unit of the Network for Environmental and Reproductive Health (EcoFoodFertility Project), "Oliveto Citra Hospital", Oliveto Citra, Salerno, Italy

3 Andrology Unit and Service of Lifestyle Medicine in UroAndrology, Local Health Authority (ASL) "St. Francis of Assisi Hospital", Oliveto Citra, Salerno, Italy

*Address all correspondence to: salvatoreraimondo57@gmail.com

IntechOpen

References

[1] Prüss-Ustün A., Wolf J., Corvalán CF., Bos R., Neira MP. Preventing disease through healthy environments: a global assessment of the burden of disease from environmental risks. 2016 World Health Organization

[2] Olea N., Fenandez MF. Chemicals in the environment and human male fertility. 2007 Occup Environ Med 64(7): 430-431

[3] Stachel B., Dougherty RC., Lahl U., Schlosser M., Zeschmar B. Toxic environmental chemicals in human semen: analytical method and case studies.1989 Andrologia;21(3):282-91

[4] Kimberley Bryon-Dodd. Environmental chemicals affect sperm epigenetics. BioNews 2017

[5] Montano L., Bergamo P., Andreassi MG., Lorenzetti S. The role of human semen as an early and reliable tool of environmental impact assessment on human health. Full Chapter in Final Book Title & ISBN: Spermatozoa – Facts and Perspectives. 2018a InTechOpen; 73231

[6] Montano L., Raimondo S., Gentile M., Notari T., Bifulco R., Porciello G., Gentile T. The role of synergic action between alpha-tocoferol and lifestyle on reduction of p53 protein in human spermatozoa (preliminary data. EcoFoodFertility). 2018b Reprod Toxicol; 80:29-30

[7] Mima, M., Greenwald, D. & Ohlander, S. Environmental Toxins and Male Fertility. 2018 Curr Urol Rep 19, 50

[8] Perheentupa A. Male infertility and environmental factors. 2019 Global Reproductive Health; 4(2) e28

[9] Sengupta P., Banerjee R. Environmental toxins: Alarming impacts of pesticides on male fertility. 2013 Human & Experimental Toxicology; 33(10): 2017-1039

[10] Tang Q., Wu W., Zhang J., Fan R., Liu Mu. Environmental factors and male Infertility 2018. DOI 10.5772 Interchopen. 71553

[11] Levine H., Jorgensen N., Martino-Andrade A., Memdiola J., Weksler-Derri D., Mindlis I., Pinotti R., Swan SH. Temporal trends in sperm count: a systematic rewiew and meta-regression analysis. 2017 Hum Reprod Update. 1; 23(6): 646-659

[12] Mishra, P., Negi, M.P.S., Srivastava, M.et al., Decline in seminal quality in Indian men over the last 37 years. 2018 Reprod Biol Endocrinol;16,103

[13] Sengupta P., Borges E. Jr, Dutta S., Krajewska-Kulak E. Decline in sperm count in European men during the past 50 years. 2018 Hum Exp Toxicol;37(3): 247-255

[14] Siqueira, S., Ropelle, A.C., Nascimento, J.A.A. et al., Changes in seminal parameters among Brazilian men between 1995 and 2018. 2020 Sci Rep;10: 6430

[15] Yuan HF., Shangguan HF., Zheng Y., Meng TQ., Xiong CL., Guan HT. Decline in semen concentration of healthy Chinese adults: evidence from 9357 participants from 2010 to 2015. 2018 Asian J Androl;20(4):379-384

[16] Levine AJ., Oren M. The first 30 years of p53: Growing ever more complex. Nat. Rev. Cancer. 2009;9:749-758

[17] Luo Q, Beaver JM, Liu Y, Zhang Z. Dynamics of p53: A master decider of cell fate. Genes (Basel). 2017;8(2):66

[18] Hanna K., Coussen C. Rebuilding the Unity of Health and Environmental: A New Vision of Environmental Health for the 21st Century. Washington, D.C.: National Academies Press, 2001

[19] Koren HS., Butler CD. The interconnection between the built environment ecology and health. In: Morel B., Linkov I. (eds) Environmental Security and Environmental Management: The Role of Risk Assessment. NATO Security through Sciences Series, vol5. Springer Dordrecht, 2006

[20] Yonish-Rouach E., Resnitzky D., Lotem J., Sachs L., Kimchi a. Oren M. wild-type p53 induces apoptosis of myeloid leukaemic cells that is inhibited by interleukin-6. 1991 Nature;352(6333): 345-347

[21] Johnson P., Chung S., Benchimol S. Growth suppression of friend virus-transformed Erythroleukemia cells by p53 protein is accompanied by Hemoglobin production and is sensitive to erythropoietin. 1993 Mol Cell Biol;Vol. 13, No. 3 1456-1463

[22] Shaw P., Bovey R., Tardy S., Sahli R., Sordat B., Costa J. Induction of apoptosis by wild-type p53 in a human colon tumor-derivedcellline 1992 Biochemistry;89:4495-4499

[23] Ramqvist T., Magnusson KP., Wang Y., Wiman KG. et al., Wild-type p53 induces apoptosis in a Burkitt lymphoma (BL) line that carries mutant p53. 1993 Oncogene;8(6):1495-500

[24] Okamura S., Arakawa H., Tanaka T., Nakanishi H., Ng CC., Taya Y., Monden M., Nakamura Y. p53DINP1, a p53-inducible gene, regulates p53-dependent apoptosis.2001 Mol. Cell; 8(1): 85-94

[25] Aubrey BJ., Kelly GL., Janic A., Herold MJ., Strasser A. How does p53 induce apoptosis and how does this relate to p53-mediated tumour suppression? 2018 Cell Death and Differentiation;25:104-113

[26] Eischen CM, Lozano G. The Mdm network and its regulation of p53 activities: A rheostat of cancer risk. Hum Mutat 2014; 35: 728-737

[27] Horn HF, Vousden KH. Coping with stress: Multiple ways to activate p53. Oncogene 2007; 26: 1306-1316

[28] Harris SL, Levine AJ. The p53 pathway: Positive and negative feedback loops.Oncogene 2005; 24: 2899-2908

[29] Oren M. Regulation of the p53 tumor suppressor protein. 1999 J Biol Chem;274: 36031-36034

[30] Appella E., Anderson CW. Post-translational modifications and activation of p53 by genotoxic stresses. 2001 Eur J Biochem; 268: 2764-2772

[31] Kubbutat MH., Vousden KH. Keeping an old friend under control: Regulation of p53 stability. 1998 Mol Med Today; 4: 250-256

[32] Ryan KM, Phillips AC, Vousden KH. Regulation and function of the p53 tumor suppressor protein. 2001 Curr Opin Cell Biol; 13: 332-337

[33] Yoshida K, Miki Y. The cell death machinery governed by the p53 tumor suppressor in response to DNA damage. Cancer Sci 2010;101(4):831-835

[34] Heller CG., Clermont Y. Spermatogenesis in man: an estimate of its duration. 1963 Science; 140 (3563): 184-186

[35] Aoki VW., Moskovtsev SI., Willis J., Liu L., Mullen JB., Carrell DT. DNA integrity is compromised in protamine-deficient human sperm. 2005 J Androl;26:741-748

[36] Aziz N., Agarwal A. Evaluation of sperm damage: Beyond the World Health Organization criteria. Fertil Steril 2008;90:484-485

[37] Kosower NS., Katayose H., Yanagimachi R. Thioldisulfide status and Acridine Orange fluorescence of mammalian sperm nuclei. 1992 J Androl;13:342-348

[38] Pivot-Pajot C., Caron C., Govin J., Vion A., Rousseaux S., Khochbin S. Acetylation-dependent chromatin reorganization by BRDT, a testis-specific bromodomain-containing protein. 2003 Mol Cell Biol;23:5354-5365

[39] Marcet-Ortega M., Pacheco S., Martínez-Marchal A., Castillo H., Flores E., Jasin M., Keeney S., Roig I. p53 and TAp63 participate in the recombination-dependent pachytene arrest in mouse spermatocytes. PLOS Genetics 2017;13(6): e1006845

[40] Bornstein C., Brosh R., Molchadsky A., et al., SPATA18, a spermatogenesis-associated gene, is a novel transcriptional target of p53 and p63. Mol Cell Biol. 2011;31:1679-1689

[41] Zalzali H., Rabeh W., Najjar O., Ammar RA., Harajly M., Saab R. Interplay between p53 and Ink4c in spermatogenesis and fertility, Cell Cycle 2018; 17(5): 643-651

[42] Toyoshima M. Analysis of p53 dependent damage response in sperm irradiated mouse embryos. 2009 J Radiat Res; 50:11-17

[43] Marty SM., Singh NP., Stebbins KE., Ann Linscombe V., Passage J., Bhaskar Gollapudi B. Initial insights regarding the role of p53 in maintaining sperm DNA integrity following treatment of mice with ethylnitrosourea or cyclo-phosphamide. 2010 Toxicol Pathol;38: 244-257

[44] Tang W., Willers H., Powell SN. p53 directly enhances rejoining of DNA double-strand breaks with cohesive ends in gamma-irradiated mouse fibroblasts. 1999 Cancer Res;59:2562-2565

[45] Riley T., Sontag E., Chen P., Levine A. Transcriptional control of human p53-regulated genes. 2008 Nat Rev Mol Cell Biol;9:402-412

[46] Nagata S. Apoptosis by death factor. Cell. 1997;88(3):355-365

[47] Gil-Guzman E., Ollero M., Lopez MC., Sharma RK., Alvarez JG., Thomas AJ Jr. Differential production of reactive oxygen species by subsets of human spermatozoa at different stages of maturation. Hum Reprod. 2001;16: 1922-1930

[48] Tejada RI., Michell JC., Norman A., Marik JJ., Friedman S. a test for the practical evaluation of male fertility by Acridine Orange (AO) fluorescence. 1984 Fertil Steril; 42(1): 87-91

[49] Penn DJ., Potts WK. The evolution of mating preferences and major histocompatibility complex genes. 1999 American Naturalist;153 (2):145-164

[50] Otts RJ., Newbury CJ., Smith G., Notarianni LJ., Jefferies TM. Sperm chromatin damage associated with male smoking. 1999 Mutat Res; 423: 103-107

[51] Traber MG., van der Vliet A., Reznick AZ., Cross CE. Tobacco-related diseases. Is there a role for antioxidant micronutrient supplementation? 2000 Clin Chest Med; 21: 173-187

[52] Kunzle R., Mueller MD., Hanggi W., Birkhauser MH., Drescher H., Bersinger NA. Semen quality of male smokers and nonsmokers in infertile couples. Fertil Steril. 2003; 79: 287-291

[53] Rodgman A., Perfetti T.O. The Chemical Components of Tobacco and Tobacco Smoke. 2013, II edition. CRC Press, Taylor and Francis Group

[54] Alchinbayev MK., Aralbayeva AN., Tuleyeva LN. Aneuploidies level in sperm nuclei in patients with infertility. Mutagenesis 2016 Sep;31(5):559-565

[55] Federico A, Morgillo F, Tuccillo C, Ciardiello F, Loguercio C. Chronic inflammation and oxidative stress in human carcinogenesis. Int J Cancer. 2007;121(11):2381-6.

[56] Aitken RJ., West KM. Analysis of the relationship between reactive oxygen

species production and leucocyte infiltration in fractions of human semen separated on Percoll gradients. Int J Androl. 1990; 13: 433-451

[57] Park J-H., Zhuang J., Li J., Hwang PM. p53 as guardian of the mitochondrial genome. 2016 FEBS Lett; 590(7): 924-934

[58] Raimondo S., Gentile T., Gentile M., et al., P53 protein evaluation on spermatozoa DNA in fertile and infertile males. J Hum Reprod Sci 2019;12(2):114-121

[59] Cocuzza M., Sikka SC., Athayde KS., Agarwal A. Clinical relevance of oxidative stress and sperm chromatin damage in male infertility: an evidence based analysis. 2007 Int Braz J Urol;33(5):603-21

[60] Lewis SE., Boyle PM., McKinney KA., Young IS., Thompson W. Total antioxidant capacity of seminal plasma is different in fertile and infertile men. Fertil Steril. 1995; 64: 868-870

[61] Sies H. Strategies of antioxidant defense. Eur J Biochem. 1993; 215(2): 213-219

[62] Huszar G., Sbracia M., Vigue L., Miller DJ., Shur BD. Sperm plasma membrane remodeling during spermiogenetic maturation in men: Relationship among plasma membrane beta 1,4-galactosyltransferase, cytoplasmic creatine phosphokinase, and creatine phosphokinase isoform ratios. Biol Reprod. 1997; 56: 1020-1024

[63] Hendin BN., Kolettis PN., Sharma RK., Thomas AJ Jr., Agarwal A. Varicocele is associated with elevated spermatozoal reactive oxygen species production and diminished seminal plasma antioxidant capacity. 1999 J Urol; 161: 1831-1834

[64] Buettner GR. The pecking order of free radicals and antioxidants: Lipid peroxidation, alpha-tocopherol, and ascorbate. Arch Biochem Biophys. 1993; 300: 535-543

[65] Halliwell B. How to characterize a biological antioxidant. Free Radic Res Commun. 1990; 9: 1-32

[66] Gomez E., Irvine DS., Aitken RJ. Evaluation of a spectrophotometric assay for the measurement of malondialdehyde and 4-hydroxyalkenals in human spermatozoa: Relationships with semen quality and sperm function. Int J Androl. 1998; 21: 81-94

[67] Bosco L., Notari T., Rovolo G., Roccheri M., Martino C., Chiappetta R., Carone D., Lo Bosco G., Carillo L., Raimondo S., Guglielmino A., Montano L. Sperm DNA fragmentation: An early and reliable marker of air pollution. Environ Toxicol Pharmacol 2018;58:243-249

[68] Kumar S. Occupational exposure associated with reproductive dysfunction. J Occup Health. 2004; 46: 1-19

[69] Gate L., Paul J, Ba GN, Tew KD, Tapiero H. Oxidative stress induced in pathologies: the role of antioxidants. 1999 Biomed Pharmacother;53(4):169-80

[70] De Rosa M., Zarrilli S., Paesano L., Carbone U., Boggia B., Petretta M., et al.,: Traffic pollutants affect fertility in men. Hum Reprod. 2003; 18(5): 1055-1061

[71] Sallmen M., Lindbohm ML., Anttila A., Taskinen H., Hemminki K. Time to pregnancy among the wives of men occupationally exposed to lead. Epidemiology. 2000a; 11: 141-147

[72] Sallmen M., Lindbohm ML., Nurminen M. Paternal exposure to lead and infertility. Epidemiology. 2000b; 11: 148-152

[73] Sokol RZ., Kraft P., Fowler IM., Mamet R., Kim E., Berhane KT. Exposure to environmental ozone

affects semen quality. Environ Health Perspect. 2006; 114: 360-365

[74] Dalvie MA., et al., The long-term effects of DDT exposure on semen, fertility and sexual function of malaria vector-control workers in Limpopo Province, South Africa. 2004 Environ Res; 96: 1-8

[75] Hauser R., Chen Z., Potheir L., Ryan L., Altshul L. The relationship between human semen parameters and environmental exposure to poly-chlorinated biphenyls and p,p'-DDE. Environ Health Perspect. 2003; 111: 1505-1511.

[76] Kruger T. et al., Xenoandrogenic activity in serum differs across Europe and Inuit populations. Environ Health Perspect. 2007; 115: 21-27

[77] Swan SH., Kruse RL., Liu F., et al., The study for future families research group. Semen quality in relation to biomarkers of pesticide exposure. 2003 Environ Health Perspect;111: 1478-1484

[78] Marty SM., Singh NP., Holsapple MP., Gollapudi BB. Influence of p53 zygosity on select sperm parameters of the mouse. 1999 Mutation Res;427:39-45

[79] Yin Y., Stahl BC., Dewolf WC., Morgentaler a. p53mediated germ cell quality control in spermatogenesis. 1998 Dev Biol;204(1): 165-171

[80] Cruz FD., Lume C., Silva JV., Nunes A., Castro I., Silva R., Silva V., Ferreira R., Fardilha M. Oxidative stress markers: can they be used to evaluate human sperm quality? 2015 Turk J Urol;41(4):198-207

[81] Samplaski MK., Dimitromanolakis A., Lo KC., Grober ED., Mullen B., Garbens A., larvi KA. The relationship between sperm viability and DNA fragmentation rates. 2015 Reprod Biol Endocrinol;13:42

[82] Fowler BA., Whittaker MH., Lipsky M., Wang G., Chen XQ. Oxidative stress induced by lead, cadmium and arsenic mixtures: 30-day, 90-day, and 180-day drinking water studies in rats: an overview. 2004 Biometal; 17(5): 567-568

[83] Raimondo S., Gentile T., Cuomo F., De Filippo S., Aprea GE., Guida J. quantitative evaluation of p53 as a new indicator of DNA damage in human spermatozoa. 2014 J Hum Reprod Sci: 7 (3); 212-217

[84] Gill P., Jeffreys AJ., Werrett DJ.: Forensic application of DNA 'fingerprints'. Nature 1985, 318:577-579

[85] Cassuto NG., Hazout A., Hammoud I., Balet R., Bouret D., Barak Y., Jellad S., Plouchart JS., Yazbeck C. Correlation between DNA defect and sperm-headmorfology (2012) ReproductiveBioMedicine Online;24:211-218

[86] Elbashir S., Magdi Y., Rashed A., Ahmedibrahim M., Edris Y, Abdelaziz AM. Relationship between sperm progressive motility and DNA integrity in fertile and infertile men. 2018 Journal Middle East Fertility Society;23(3):195-198

[87] Aitken RJ., De Iuliis GN., Finnie JM., Hedges A., McLachlan RI. Analysis of the relationships between oxidative stress, DNA damage and sperm vitality in a patient population: development of diagnostic criteria. (2010) Hum Reprod; 25(10): 2415-2426

[88] Micic S., Lalic N., Djordjevic D., Bojanic N., Virmani A., Agarwal A. Sperm vitality and DNA fraagmentation index (DFI) are good predictors of progressive sperm motility in oligozooasthenospermic men treated with metabolic abd essential nutrients. 2019 Hum Reprod; 33(3): 149-

[89] Agarwal A., Sharma R., Harlev A., Esteves SC. Effect of varicocele on

semen characteristics according to the new 210 WHO criteria: A systematic rewiew and meta-analysis. 2016 Asian J Androl;18:163-170

[90] Aitken RJ., Fisher HM., Fulton N., Gomez E., Knox W., Lewis B., Irvine S. reactive oxygen species generation by human spermatozoa is induced by exogenous NADPH and inhibited by the flavoprotein inhibitors diphenylene iodonium and quinacrine. 1997 Mol Reprod Dev;47(4):468-482

[91] Mortimer D. Practical Laboratory Andrology. New York (USA): Oxford University Press.1994

[92] Griswold MD. Spermatogenesis: the commitment to meiosis. 2016 Physiol Rev; 96(1): 1-17

[93] Junqueira LC., Carneiro J. Compendio di istologia. 2006; Padova, Piccin V edizione

[94] Gonzalez-Mariscal L., Quiros M., Diaz-Coranguez M., Bautista P. Tight Junctions, Current Frontiers and perspectives in cell Biology. 2012: Edited by Stevo Najman, IntechOpen

[95] Mruk DD., Cheng CY. The mammalian blood-testis barrier: its biology and regulation. 2015 Endocrine Reviews;36(5):564-591

[96] Phillips BT., Gassei K., Orwig KE., Spermatogonial stem cell regulation and spermatogenesis. 2010 Phil Trans R Soc B; 365(1546): 1663– 1678

[97] Gu, J., Chen D., Rosenblum J., Rubin R., and Yuan Z. M. Identification of a sequence element from p53 that signals for Mdm2-targeted degradation. 2000 Mol Cell Biol; 20: 1243-1253

[98] Rahman S., Lee JS., Kwon WS., Pang MG. Sperm proteomics: road to male fertility and contraception. 2013 Int J Endocrinol Article; ID 360986, 11pages

[99] Gebel J., Tuppi M., Krauskopf K., Coutandin D., Pitzius S., Kehrloesser S., Osterburg C., Dötsch V. Control mechanisms in germ cells mediated by p53 family proteins. 2017 Journal of Cell Science;130:2663-2267

[100] Gunes S., Al-Sadaan M., Agarwal A. Spermatogenesis, DNA damage and DNA repair mechanisms in male infertility. 2015 Reproductive BioMedicine Online;31:309-319

[101] Lane DP. Cancer. P53 guardian of the genome. 1992 Nature; 358:15-16

[102] Rotter, V., Schwartz, D., Almon, E., Goldfinger, N., Kapon, A., Meshorer, A., Donehower, L.A., Levine, a.J. mice with reduced levels of p53 exhibit the testicular giant cell degenerative syndrome. 1993 PNAS; 90 (19): 9075-9079

[103] Chresta, C.M., Masters, J R. W., and Hickman, J.A. Hypersensitivity of human testicular tumors to etoposide-induced apoptosis is associated with functional p53 and a high bax: bcl-2 ratio. 1996 Cancer Res; 53: 1834-41

[104] Masters J.R.W., Osborne, E.J., Walker, M.C., and Parris, C. N. Hypersensitivity of human testis-tumor cell lines to chemotherapeutic drugs. 1993 Int. J. Cancer; 53:340-346

[105] Socher SA., Yin Y., Dewolf, WC., Morgentaler, A. Temperature-mediated germ cell loss in the testis is associated with altered expression of the cell-cycle regulator p53. 1997 J Urol; 157: 1986-1989

[106] Olivier R. T. Testicular cancer. 1996 Curr Opin Oncol; 8: 252-258

[107] Olivier, R. T. Testicular cancer. 1997 Curr Opin Oncol; 9: 287-294

[108] Zamble, D. B., Jacks, T., and Lippard, S.J. p53dependent and independent responses to cisplatin in mouse testicular teratocarcinoma cells. 1998 Proc. Natl Acad Sci; 95: 6163-6168

[109] Lutzker, S. G., and Levine, A. J. A functionally inactive p53 protein in teratocarcinoma cells is activated by either DNA damage or cellular differentiation. 1996 Nat Med; 2: 804-810

[110] Gersten, K. M., and Kemp, C. J. Normal meiotic recombination in p53-deficient mice. 1997 Nat Genet; 17:378-379

[111] Odorisio, T., Rodriguez TA., Evans EP., Clarke AR., Burgoyne PS. The meiotic checkpoint monitoring synapsis eliminates spermatocytes via p53-indipendent apoptosis. 1998 Nat Genet; 18: 257-261

[112] Katz DF., Overstreet JW. Sperm motility assessmente by videomicrografy. 1981 Fert Steril;35(2):188-193

[113] Canale D., Giorgi PM., Gasperini M., Pucci E., Barletta D., Gasperi M., Martino E. Inter and Intra-Individual Variability of Sperm Morphology After Selection With Three Different Techniques: Layering, Swimup From Pellet and Percoll. 1994 J. Endocrinol Invest; 17(9):729-32

[114] Pinto S., Carrageta DF., Alves MG., Rocha A., Agarwal A., Barros A., Oliveira P. Sperma selection strategies and their impact on assisted reproductive technology outcomes. 2020 Andrologia; 00:e13725

[115] Allahbadia GN., Gandhi G. et al. The art & Science of Assisted Reproductive Techniques (ART) Medical Ltd. 2017

[116] Anifandis G., Bounartzi C, Messini I, Dafopulos K, Sotiriou S, Messinis E. The impact of cigarette smoking and alcohol consumption on sperm parameters and sperm DNA fragmentation (sDF) measured by Halosperm. 2014 Archives of Gynecology and Obstetric, 290(4):777-782

[117] Kirsty A., Nisenblat V, Norman R. Lifestyle factors in people seeking infertility treatment – A review.2010 Australian and New Zealand Journal of Obstetric and Gynecology;50(1):8-20

[118] La Vignera S., Condorelli R, Vicari E, D'Agata R, Calogero AE. Effects of varicocelectomy on sperm DNA fragmentation, mitochondrial function, chromatin condensation, and apoptosis. 2012 J Androl; 33:389-396

[119] Smit M., Romijn JC, Wildhagen MF, Veldhoven JL, Weber RF, Dohle GR. Decreased sperm DNA fragmentation after surgical varicocelectomy is associated with increased pregnancy rate. 2010 J Urol;183:270-274

[120] Wang YJ., Zhang RQ, Lin YJ, Zhang RG, Zhang WL. Relationship between and sperm DNA damage and the effect of varicocele repair: a meta-analysis. 2012 Reprod Biomed online; 25(3): 307-314

[121] Alahmar AT. Sperm motility using direct swim-up technique at two different temperatures. 2018 Gynecol Obstrt; 8:86

[122] Boomsma H. Semen preparation techniques for intrauterine insemination. Cochrane Database of Systematic Reviews. 2004

[123] Aitken RJ., Clarkson JS. Significance of reactive oxygen species ant antioxidants in defining the efficacy of sperm preparation techniques. 1988 J Androl; 9:367-376

[124] Volpes A., Sammartano F., Rizzari S., Gullo S., Marino A., Allegra A. The pellet swim-up is the best technique for sperm preparation during in vitro fertilization procedures. 2016 J Assist Reprod Genet; 33(6): 765-770

[125] Morshedi M. Efficacy and pregnancy outcome of two methods of semen preparation for intrauterine insemination: a prospective randomized study. 2003 Fert Steril;79 (3):1625-1632

[126] Xue X., Wang WS, Shi JZ, Zhang SL, Zhao WQ, Shi WH, Guo BZ, Qin Z. efficacy of swim-up versus density gradient centrifugation in improving sperm deformity rate and DNA fragmentation index in semen samples from teratozoospermic patients. 2014 J Assist Reprod Genet; 31(9):1161-1166

[127] Marchetti C., Gallego M.A., Deffosez A., Formstecher P.,Marchetti P. Staining of human sperm with fluorochrome-labeled inhibitor of caspases to detect activated caspases: Correlation with apoptosis and sperm parameters. 2004 Hum Reprod;19: 1127-1134

[128] Said T., Agarwal T., Grunewald S., Rasch M., Baumann T. and Kriegel C. et al. Selection of nonapoptotic spermatozoa as a new tool for enhancing assisted reproduction outcomes: an in vitro model 2006 Biol Reprod; 74: 530-537

[129] Raimondo S., Gentile T., Gentile M., Donnarumma F., Esposito G., Morelli A., De Filippo S., Cuomo F. comparing different sperm separation techniques for ART, through quantitative evaluation of p53 protein. 2020 J Hum Reprod Sci; 13(2): 117-124

[130] Levine AJ., Tomasini R., McKeon FD., Mak TW. The p53 family: guardians of maternal reproduction. 2011 Nat Rev Mol Cell Biol, 12(4), 259-265

[131] Chandrakanthan V., Li A., Chami O., O'Neill C. Effects of in vitro fertilization and embryo culture on TRP53 and Bax expression in B6 mouse embryos. 2006 Reprod Biol Endocrinol;4, 61-70

[132] Li A., Chandrakanthan V., Chami O., O'Neill C. culture of zygotes increases Trp53 [corrected] expression in B6 mouse embryos, which reduces embryo viability. 2007 Biol Reprod; 76(3): 362-367

[133] Hanson HA., Mayer EN., Anderson RE., Aston KI., Carrel DT., Berger J., Lowrance WT., Smith KR., Hotaling JM. Risk of childhood mortality in family members of men with poor semen quality. 2017 Hum Reprod; 22(1) 239-247

[134] Raimondo S., Notari T., Gentile M., Cirmeni M., Gentile T., Montano L., Borsellino G. level of p53 protein in human spermatozoa, embryo quality and pregnancy rate. EcoFoodFertility project (preliminary data). 2018 Hum Reprod;22:85

Chapter 6

A Study of p53 Action on DNA at the Single Molecule Level

Kiyoto Kamagata

Abstract

The transcription factor p53 searches for and binds to target sequences within long genomic DNA, to regulate downstream gene expression. p53 possesses multiple disordered and DNA-binding domains, which are frequently observed in DNA-binding proteins. Owing to these properties, p53 is used as a model protein for target search studies. It counters cell stress by utilizing a facilitated diffusion mechanism that combines 3D diffusion in solution, 1D sliding along DNA, hopping/jumping along DNA, and intersegmental transfer between two DNAs. Single-molecule fluorescence microscopy has been used to characterize individual motions of p53 in detail. In addition, a biophysical study has revealed that p53 forms liquid-like droplets involving the functional switch. In this chapter, the target search and regulation of p53 are discussed in terms of dynamic properties.

Keywords: p53, single molecule, fluorescence, DNA, disordered, diffusion, jumping, intersegmental transfer, sliding, hopping, target search, liquid–liquid phase separation

1. Introduction

p53 is a multifunctional transcription factor that induces cell cycle arrest, DNA repair, and apoptosis, thereby suppressing cell cancerization [1, 2]. It is referred to as a guardian of the genome that determines cell fate. When p53 is activated by various stress factors, it searches for and binds to target DNA sequences and regulates the expression of downstream genes. p53 is composed of an N-terminal (NT) domain, core domain, linker, tetramerization (Tet) domain, and C-terminal (CT) domain. The core and Tet domains possess specifically folded structures, while other domains are intrinsically disordered [3–5]. p53 forms a tetramer via Tet domains [5]. Core and CT domains are involved in its binding to DNA sequences in a specific and nonspecific manner, respectively [6]. Fifty percent of gene mutations in tumor cells were found in p53, and many of the identified mutations were located in structured domains, which inhibited target DNA binding [3]. Comprehensive mutagenesis analysis supports the correlation between the structured domains of p53 and its function [7]. Since p53 possesses common properties frequently observed in DNA-binding proteins, including oligomerization, disordered regions, and multiple DNA-binding domains [8], it is used as a model protein in the target search study described below [9–11].

The target DNAs for p53 were ~ 20 bp, while the genomic DNA was ~10^9 bp. Accordingly, p53 was required to search for small targets efficiently from within vast lengths of non-target DNAs. This is known as a target search problem for

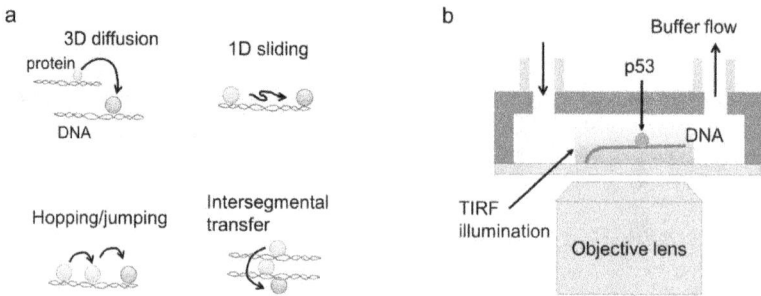

Figure 1.
Target search dynamics of DNA-binding proteins and visualization of p53 dynamics on DNA by single-molecule fluorescence microscopy. (a) Schematic diagram of four target search dynamics. (b) Schematic diagram of single-molecule fluorescence microscope and flow cell. In the flow cell, one end of the DNA is tethered to the surface and it is stretched by buffer flow. p53 molecule labeled to a fluorescence dye is illuminated by TIRF and the fluorescence is detected by EM-CCD through an objective lens. Panels (a) and (b) are adapted from ref. [12] and ref. [13] with some modifications, respectively.

sequence-specific DNA-binding proteins. To solve this problem, a facilitated diffusion mechanism has been proposed for DNA-binding proteins. The facilitated diffusion is the integration of three-dimensional (3D) diffusion in solution, one-dimensional (1D) diffusion along DNA, hopping/jumping along DNA, and intersegmental transfer between two DNAs (**Figure 1a**). In 3D diffusion, p53 diffuses in solution, altering the search sites on genomic DNA. In 1D sliding, it moves along the DNA, while maintaining continuous contact. In addition, p53 hops or jumps along DNA (within 100 bp of jump). Intersegmental transfer enables p53 to move from one DNA to another without dissociation. Theoretical studies suggest that the integration of multiple search dynamics, while not requiring all dynamics, can facilitate the target search [14–17]. The facilitation factor depends on various physical parameters, such as diffusion coefficient along DNA, residence time on DNA, dissociation time in solution, and frequency of transfer and jump.

How does p53 solve the target search problem using facilitated diffusion? How is the target search and binding of p53 regulated? In this chapter, I explain the facilitated diffusion and regulation of p53 based on recently accumulated single-molecule data.

2. Single-molecule fluorescence microscopy

Single-molecule fluorescence microscopy enables the differentiation and characterization of individual search dynamics of DNA-binding proteins, including p53, as reported previously [18–24]. In general, the system combines a fluorescence microscope and a flow cell (**Figure 1b**). In the flow cell, one end of the DNA is tethered to the surface, and it is stretched by buffer flow. Several methods have been proposed for tethering DNAs [18, 25–29]. For example, a DNA garden is a simple method for producing DNA arrays, in which neutravidin molecules are printed in a line on polymer-coated coverslips, and biotinylated DNAs are tethered to the printed neutravidin [29]. p53 molecules labeled with a fluorescence dye are introduced into the flow cell using a syringe pump. The fluorescent p53 bound to DNA is selectively illuminated by total internal reflection fluorescence (TIRF). p53 molecules on DNA are detected as fluorescent spots on the sequential images of an electron-multiplying charge-coupled device (EM-CCD). The positions of molecules were tracked using an appropriate analysis program to visualize the search dynamics of p53.

3. Target search dynamics of p53

In 2008, 1D sliding of p53 along DNA was observed for the first time using single-molecule fluorescence microscopy [30]. This observation was consistent with a reported indirect evidence that p53 dissociated rapidly from short DNA in the absence of blocks at ends by sliding off from DNA [31]. In 2011, a study of p53 mutants deleting either of two DNA-binding proteins revealed that p53 can slide along DNA using disordered CT domains [32]. This is consistent with the fact that a designed peptide targeting CT domains suppressed the 1D sliding of p53 [33]. Furthermore, 1D sliding of p53 was supported by molecular dynamics simulations [34, 35]. In 2012, it was shown that 1D sliding dynamics of p53 depends slightly on DNA sequence, suggesting that p53 feels the energy landscape based on DNA sequence through interactions between core domains and DNA [36]. In 2015, a detailed analysis of 1D sliding dynamics demonstrated that p53 possesses two sliding modes on non-target DNA [37, 38]. In the fast mode, it interacts with DNA loosely using CT domains. In contrast, in the slow mode, it binds tightly to DNA using core and CT domains (**Figure 2a**). In 2017, the disordered linker was

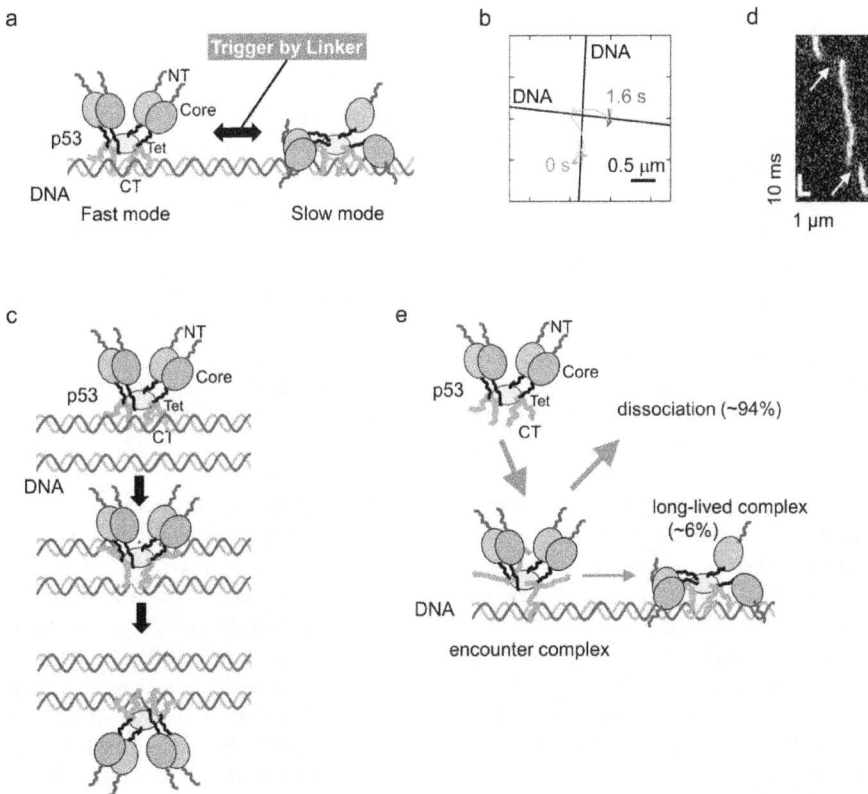

Figure 2.
Target search dynamics of p53. (a) Schematic diagram of two modes for 1D sliding p53 along DNA. p53 is composed of the NT (purple), Core (green), linker (black), Tet (yellow), and CT (pink) domains. The switch between two modes is triggered by the linker. (b) Typical single-molecule data showing intersegmental transfer of p53 between crisscrossing DNAs. (c) Schematic diagram of intersegmental transfer of p53 between two DNAs. p53 uses CT domains (pink) for the transfer. (d) Typical single-molecule data showing jumping of p53 along DNA (white traces). Arrows denote the jumping events. (e) Schematic diagram of encounter complex formation of p53 and conversion from the encounter complex to long-lived complex. Panel (b) is adapted from ref. [39] with some modifications. Panels (d) and (e) are adapted from ref. [40] with some modifications.

identified to trigger the switch between the two modes (**Figure 2a**) [41]. In 2016, the target recognition process of p53 was characterized in detail [42]. The results demonstrated that target recognition occurs mainly via 1D sliding. The target recognition of p53 was quite low (the successful recognition probability was 7%), but it was enhanced two-fold upon a post-translational modification. Accordingly, 1D sliding is considered as one of the important dynamics in the target search and binding of p53.

In 2018, intersegmental transfer of p53 was examined using ensemble kinetic and single-molecule fluorescence measurements [39]. After the solutions of p53 bound to fluorescently labeled DNA and non-labeled DNA were mixed, the transfer reaction of p53 was monitored between the two DNAs. The observed reactions included the dissociation of p53 from one DNA and its transfer to the other. Actually, as the concentration of non-labeled DNA increased, the observed rate constant increased, suggesting intersegmental transfer. The rate constant of the transfer was ~10^8 M^{-1} s^{-1}, which is close to the diffusion limit. Furthermore, single-molecule tracking of p53 on crisscrossed DNAs demonstrated that p53 moves along the first DNA and then moves along the second DNA through the transfer at the intersection (**Figure 2b**). A study of p53 mutants deleting either of two DNA-binding domains identified that p53 binds to the first DNA and then to the second DNA using disordered CT domains at the same time; it then releases the first DNA, resulting in a transfer between the two DNAs (**Figure 2c**). This mechanism is supported by molecular dynamics simulations of p53 [43].

In 2020, the hopping/jumping of p53 on DNA was investigated [40]. Hopping/jumping was expected to occur at a time scale that is faster than the time resolution of the microscope (ex. 33 ms). To detect these events, the time resolution of the microscope was improved to 500 μs by optimizing the fluorescence excitation based on critical angle TIRF illumination and by utilizing the time delay integration mode of the EM-CCD [40]. Using the sub-millisecond-resolved microscope, jumping events of p53 along DNA were directly detected (arrows in **Figure 2d**). The jump frequency of p53 was ~6 s^{-1}, and the jump time was 2.2 ms. Based on the study of p53 mutants deleting either of two DNA-binding domains, disordered CT domains were identified to be indispensable for the jumping of p53 along DNA [13]. Furthermore, 1D diffusion along DNA was enhanced upon increasing the salt concentration, suggesting that p53 moves along DNA by hopping DNA-binding domains. Thus, it was revealed that p53 possesses hopping and jumping dynamics along DNA.

In 2016, 3D diffusion of p53 was characterized using ensemble kinetic measurements [42]. Association rate constants for target and non-target DNAs were determined to be ~10^9 M^{-1} s^{-1}, comparable to the diffusion limits. The difference in affinity for target and non-target DNAs was attributed to the dissociation rate constants. In 2020, the association process of p53 with non-target DNA was further investigated at the single-molecule level using a sub-millisecond resolved fluorescence microscope [40]. Kymographs demonstrated that short-lived traces of p53 with an average residence time of 2.8 ms were detected in addition to long-lived traces moving along DNA. The short-lived complex was interpreted as an encounter complex. Disordered CT domains of p53 were identified to participate in the transient complex formation and in the conversion from the transient complex to the long-lived complex [13] (**Figure 2e**). The long-lived complex was further stabilized by core domains [13].

Overall, single-molecule fluorescence microscopy revealed that p53 possesses all four search dynamics proposed theoretically: 3D diffusion, 1D sliding, hopping/jumping, and intersegmental transfer. The unique structure of p53, which is a tetramer of two DNA-binding domains, enables these search dynamics. This is the first study to examine all search possibilities for a single model protein.

4. Liquid–liquid phase separation of p53

Target search and binding of p53 might be regulated by a liquid-like assembly of p53 molecules. In a liquid–liquid phase separation (LLPS), p53 molecules, which disperse in the bulk phase, assemble and form a condensed phase called liquid droplets. In the droplet phase, p53 can move fluidly while maintaining a high concentration. This fluid property in the condensed phase differs from the solid aggregation that causes malfunction of p53. Early *in vivo* studies demonstrated that p53 is recruited into cellular droplets such as Cajal and promyelocytic leukemia protein (PML) bodies [44–46]. These facts suggest that LLPS might be involved in the cellular functions of p53.

In 2020, this possibility was extensively examined using *in vitro* measurements such as scattering, DIC microscopy, and fluorescence microscopy [47]. p53 formed micrometer-sized droplets at neutral and slightly acidic pH and low salt concentrations. The fusion events of at least two droplets into a single large droplet were observed, confirming the fluidity of p53 inside the droplets (**Figure 3a**). High fluorescence intensity was detected in the droplets of p53 labeled with a fluorescent dye, supporting the high concentration of p53 in the droplets (**Figure 3b**). The droplet formation of p53 was affected by pH and salt concentrations. This suggests that attractive electrostatic interactions among local parts of p53 and repulsive net charges among whole molecules of p53 are balanced, resulting in droplet formation. Deletion of either of the disordered NT and CT domains suppressed the droplet formation of p53. This suggests multivalent electrostatic interactions between the oppositely charged NT and CT domains in p53 droplets.

The structural properties of p53 in solution and in droplet form were investigated using fluorescence resonance energy transfer (FRET) between two fluorophores labeled at two residues of p53. Since FRET depends on the distance

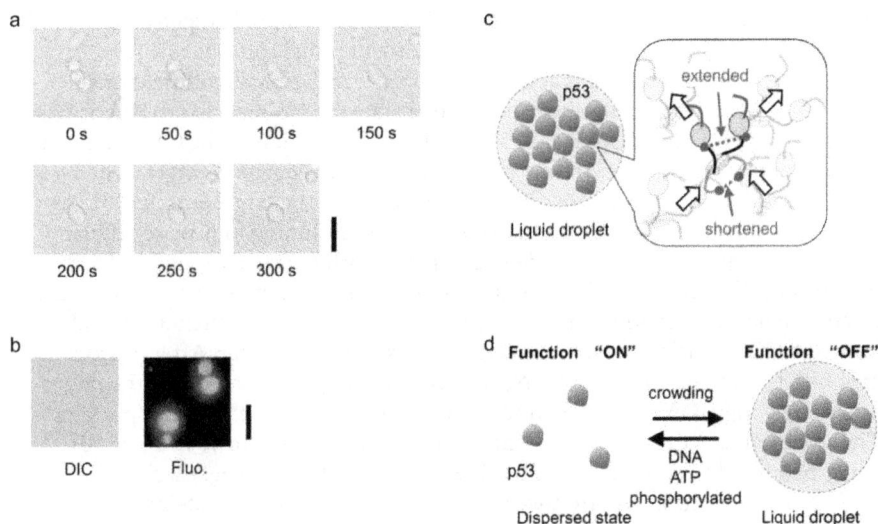

Figure 3.
Liquid droplet formation of p53 regulates its function. (a) Time course of a typical fusion event of three p53 droplets into a single droplet using DIC microscopy. (b) DIC and fluorescence images of the droplets of Alexa488-labeled p53 and non-labeled p53. Scale bars in panels (a) and (b) represent 10 μm. (c) Schematic diagram of p53 conformation in the droplet. p53 is composed of the NT (purple), Core (orange), Tet (yellow), and CT (red) domains. In the droplets, the NT and CT domains interact electrostatically. Arrows denote the structural changes on the different domains of p53 that are induced by the intermolecular interactions in a droplet. The dimer structure is displayed for clarity. (d) Functional switch model of p53. The panels (a)-(d) are adapted from ref. [47] with some modifications.

between the two fluorophores, it was used to measure the conformational changes. The distance between the core domains of p53 was slightly longer in the droplets, while the distance between the CT domains became slightly shorter (**Figure 3c**). Accordingly, p53 adopted a new tertiary structure, forming interactions with the adjacent molecules in the droplets.

Does p53 maintain binding to the target DNA after experiencing the droplet formation? The reactions of p53 binding to the target DNA were similar before and after the droplet formation. These results indicate that droplet formation of p53 is reversible, and p53 dispersed in solution from the droplets retains its DNA binding ability.

Droplet formation of p53 was found to be regulated by molecular crowding, endogenous molecules, and post-translational modification. Molecular crowding agents, mimicking the cellular crowding condition, promoted droplet formation. In contrast, ssDNA, dsDNA, and ATP suppressed it. The p53 mutant mimicking post-translational phosphorylation did not form droplets. Based on these results, a functional switch model was proposed (**Figure 3d**). Under normal cell conditions, the compartmentalization of p53 into the droplets suppresses its function as a transcriptional regulator. Under stress conditions, the activation of p53, triggered by posttranslational phosphorylation, releases p53 from the droplets and promotes target search and binding.

5. Target search and regulation model of p53

In this section, the current model of p53 is described in terms of target search and regulation. p53 functions as a transcription factor that responds to various emergency situations in cells. Under normal cell conditions, p53 turns off through the following mechanisms. First, the copy number of p53 is maintained at a low level, allowing dimers with a low affinity to target DNAs in an oligomeric state [48, 49]. Second, post-translational modifications for activating p53 are not added, for example, suppressing the target recognition of p53 [42]. Third, p53 is stored in liquid droplets [47]. These actions of p53 prevent its malfunction under normal conditions.

Under cellular stress, p53 is activated by post-translational modifications [1, 2, 50–56] and by a change in its oligomeric state from dimers to tetramers, with a high affinity for target DNAs [48, 57–59]. Phosphorylation of the CT domain of p53 triggers its release from the droplets, allowing it to engage in target search [47]. The increase in the copy number of p53 also facilitates the target search [60]. As explained above, p53 utilizes facilitated diffusion combining four search dynamics. Using 3D diffusion, p53 associates randomly with nonspecific sites of DNA, followed by dissociation. Until p53 associates with the target sequence by chance, it repeats such association and dissociation motions. If the search motion of p53 is limited to 3D diffusion, it would be a time-consuming endeavor. After p53 associates with the nonspecific site of DNA by 3D diffusion, it can search for the target sequence along DNA near the bound site using 1D sliding and hopping/jumping. The search distance of p53 per association event is estimated to be 700 bp [40], corresponding to approximately 35-fold facilitation of the target search.

In cells, genomic DNAs are covered by many DNA-binding proteins, including histones and other nucleoid proteins. These DNA-binding proteins may act as obstacles in the target search of p53. For example, when the sliding p53 collides with other DNA-binding proteins on DNA, it may not be able to bypass these obstacles due to steric hindrance, thereby limiting the search distance on DNA. However, p53 possesses two bypass mechanisms: the jumping along DNA [40] and the intersegmental transfer between two DNAs [39]. Using these motions, it can overcome such obstacles and continue its search for targets in cells. Overall, the search and regulation strategies of p53 could satisfy various cellular requirements.

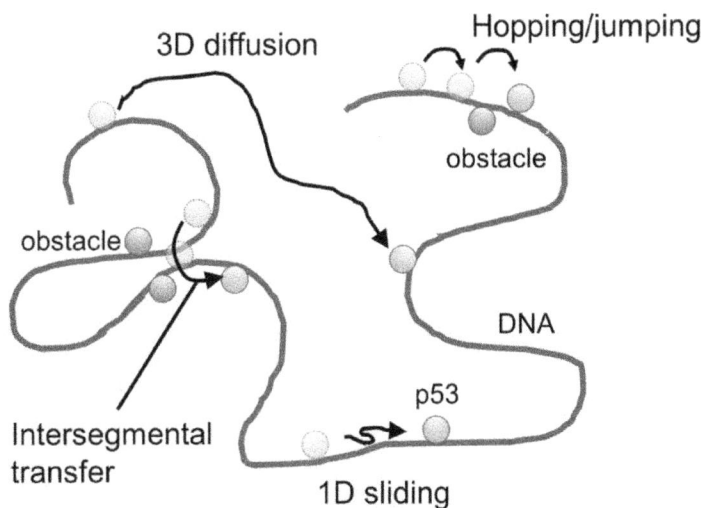

Figure 4.
Schematic diagram of target search of p53. Pink and gray circles are p53 and obstacle bound to DNA, respectively. Four search mechanisms are illustrated.

6. Conclusions

The target search and binding of p53 and its regulation have been characterized using single-molecule fluorescence microscopy and relevant biophysical methods. The accumulated data demonstrate that p53 searches for target DNAs utilizing four search dynamics: 3D diffusion in solution, 1D sliding along DNA, hopping/jumping along DNA, and intersegmental transfer between two DNAs (**Figure 4**). Especially, hopping/jumping and intersegmental transfer between two DNAs are required to bypass obstacles bound to DNA. It was reported that other DNA-binding protein with a disordered DNA-binding domain bypasses obstacles through obstacle-unbound region of DNA [24]. Since p53 possesses a similar disordered DNA-binding domain, it is not surprising that p53 possesses this bypass mechanism. Target search and binding are regulated by copy number, post-translational modifications, and liquid droplet formation. Considering that p53 can interact with many partner proteins, the partner proteins may affect the target search. Complexity in the target search and regulation of p53 would enable a response to various emergency situations in cells and be required to satisfy various cellular requirements.

Acknowledgements

I thank the corroborators for their helpful discussions on our studies of target search by p53.

Author details

Kiyoto Kamagata
Institute of Multidisciplinary Research for Advanced Materials, Tohoku University, Sendai, Japan

*Address all correspondence to: kiyoto.kamagata.e8@tohoku.ac.jp

IntechOpen

References

[1] Bieging KT, Mello SS, Attardi LD. Unravelling mechanisms of p53-mediated tumour suppression. Nature Reviews Cancer. 2014:14:359-370. DOI: 10.1038/nrc3711

[2] Beckerman R, Prives C. Transcriptional regulation by p53. Cold Spring Harbor Perspectives in Biology. 2010:2:a000935. DOI: 10.1101/cshperspect.a000935

[3] Joerger AC, Fersht AR. The tumor suppressor p53: from structures to drug discovery. Cold Spring Harbor Perspectives in Biology. 2010:2:a000919. DOI: 10.1101/cshperspect.a000919

[4] Laptenko O, Tong DR, Manfredi J, Prives C. The Tail That Wags the Dog: How the Disordered C-Terminal Domain Controls the Transcriptional Activities of the p53 Tumor-Suppressor Protein. Trends in Biochemical Sciences. 2016:41:1022-1034. DOI: 10.1016/j.tibs.2016.08.011

[5] Kamada R, Toguchi Y, Nomura T, Imagawa T, Sakaguchi K. Tetramer formation of tumor suppressor protein p53: Structure, function, and applications. Biopolymers. 2016:106:598-612. DOI: 10.1002/bip.22772

[6] Anderson ME, Woelker B, Reed M, Wang P, Tegtmeyer P. Reciprocal interference between the sequence-specific core and nonspecific C-terminal DNA binding domains of p53: Implications for regulation. Molecular and Cellular Biology. 1997:17:6255-6264. DOI: 10.1128/MCB.17.11.6255

[7] Kato S, Han SY, Liu W, Otsuka K, Shibata H, Kanamaru R, Ishioka C. Understanding the function-structure and function-mutation relationships of p53 tumor suppressor protein by high-resolution missense mutation analysis. Proceedings of the National Academy of Sciences of the United States of America. 2003:100:8424-8429. DOI: 10.1073/pnas.1431692100

[8] Vuzman D, Levy Y. Intrinsically disordered regions as affinity tuners in protein-DNA interactions. Molecular Biosystems. 2012:8:47-57. DOI: 10.1039/c1mb05273j

[9] Tafvizi A, Mirny LA, van Oijen AM. Dancing on DNA: kinetic aspects of search processes on DNA. Chemphyschem. 2011:12:1481-1489. DOI: 10.1002/cphc.201100112

[10] Kamagata K, Murata A, Itoh Y, Takahashi S. Characterization of facilitated diffusion of tumor suppressor p53 along DNA using single-molecule fluorescence imaging. J Photochem Photobiol C Photochem Reviews. 2017:30:36-50. DOI: 10.1016/j.jphotochemrev.2017.01.004

[11] Kamagata K, Itoh Y, Subekti DRG. How p53 molecules solve the target DNA search problem: a review. International Journal of Molecular Sciences. 2020:21. DOI: 10.3390/ijms21031031

[12] Kamagata K. Front line of DNA-binding protein research (Japanese). Precision Medicine. 2020:3:542-548. DOI:

[13] Subekti DRG, Kamagata K. The disordered DNA-binding domain of p53 is indispensable for forming an encounter complex to and jumping along DNA. Biochemical and Biophysical Research Communications. 2021:534:21-26. DOI: 10.1016/j.bbrc.2020.12.006

[14] Sheinman M, Kafri Y. The effects of intersegmental transfers on target location by proteins. Physical Biology. 2009:6:016003. DOI: 10.1088/1478-3975/6/1/016003

[15] Bauer M, Metzler R. Generalized facilitated diffusion model for DNA-binding proteins with search and recognition states. Biophysical Journal. 2012:102:2321-2330. DOI: 10.1016/j.bpj.2012.04.008

[16] Veksler A, Kolomeisky AB. Speed-Selectivity Paradox in the Protein Search for Targets on DNA: Is It Real or Not?, Journal of Physical Chemistry B. 2013:117:12695-12701. DOI: 10.1021/jp311466f

[17] Mahmutovic A, Berg OG, Elf J. What matters for lac repressor search in vivo--sliding, hopping, intersegment transfer, crowding on DNA or recognition?, Nucleic Acids Research. 2015:43:3454-3464. DOI: 10.1093/nar/gkv207

[18] Kabata H, Kurosawa O, Arai I, Washizu M, Margarson SA, Glass RE, Shimamoto N. Visualization of single molecules of RNA-polymerase sliding along DNA. Science. 1993:262:1561-1563

[19] Blainey PC, Luo G, Kou SC, Mangel WF, Verdine GL, Bagchi B, Xie XS. Nonspecifically bound proteins spin while diffusing along DNA. Nature Structural & Molecular Biology. 2009:16:1224-1229. DOI: 10.1038/nsmb.1716

[20] Visnapuu ML, Greene EC. Single-molecule imaging of DNA curtains reveals intrinsic energy landscapes for nucleosome deposition. Nature Structural & Molecular Biology. 2009:16:1056-1062. DOI: 10.1038/nsmb.1655

[21] Sternberg SH, Redding S, Jinek M, Greene EC, Doudna JA. DNA interrogation by the CRISPR RNA-guided endonuclease Cas9. Nature. 2014:507:62-67. DOI: 10.1038/nature13011

[22] Nelson SR, Dunn AR, Kathe SD, Warshaw DM, Wallace SS. Two glycosylase families diffusively scan DNA using a wedge residue to probe for and identify oxidatively damaged bases. Proceedings of the National Academy of Sciences of the United States of America. 2014:111:E2091–E2099. DOI: 10.1073/pnas.1400386111

[23] Kamagata K, Mano E, Ouchi K, Kanbayashi S, Johnson RC. High free-energy barrier of 1D diffusion along DNA by architectural DNA-binding proteins. Journal of Molecular Biology. 2018:430:655-667. DOI: 10.1016/j.jmb.2018.01.001

[24] Kamagata K, Ouchi K, Tan C, Mano E, Mandali S, Wu Y, Takada S, Takahashi S, Johnson RC. The HMGB chromatin protein Nhp6A can bypass obstacles when traveling on DNA. Nucleic Acids Research. 2020:48:10820-10831. DOI: 10.1093/nar/gkaa799

[25] Harada Y, Funatsu T, Murakami K, Nonoyama Y, Ishihama A, Yanagida T. Single-molecule imaging of RNA polymerase-DNA interactions in real time. Biophysical Journal. 1999:76:709-715. DOI: 10.1016/S0006-3495(99)77237-1

[26] Fazio T, Visnapuu ML, Wind S, Greene EC. DNA curtains and nanoscale curtain rods: high-throughput tools for single molecule imaging. Langmuir. 2008:24:10524-10531. DOI: 10.1021/la801762h

[27] Visnapuu ML, Fazio T, Wind S, Greene EC. Parallel arrays of geometric nanowells for assembling curtains of DNA with controlled lateral dispersion. Langmuir. 2008:24:11293-11299. DOI: 10.1021/la8017634

[28] Hughes CD, Wang H, Ghodke H, Simons M, Towheed A, Peng Y, Van Houten B, Kad NM. Real-time single-molecule imaging reveals a direct interaction between UvrC and UvrB on DNA tightropes. Nucleic Acids

Research. 2013:41:4901-4912. DOI: 10.1093/nar/gkt177

[29] Igarashi C, Murata A, Itoh Y, Subekti DRG, Takahashi S, Kamagata K. DNA garden: a simple method for producing arrays of stretchable DNA for single-molecule fluorescence imaging of DNA-binding proteins. Bulletin of the Chemical Society of Japan. 2017:90:34-43. DOI: 10.1246/bcsj.20160298

[30] Tafvizi A, Huang F, Leith JS, Fersht AR, Mirny LA, van Oijen AM. Tumor suppressor p53 slides on DNA with low friction and high stability. Biophysical Journal. 2008:95:L01–L03. DOI: 10.1529/biophysj.108.134122

[31] McKinney K, Mattia M, Gottifredi V, Prives C. p53 linear diffusion along DNA requires its C terminus. Molecular Cell. 2004:16:413-424. DOI: 10.1016/j.molcel.2004.09.032

[32] Tafvizi A, Huang F, Fersht AR, Mirny LA, van Oijen AM. A single-molecule characterization of p53 search on DNA. Proceedings of the National Academy of Sciences of the United States of America. 2011:108:563-568. DOI: 10.1073/pnas.1016020107

[33] Kamagata K, Mano E, Itoh Y, Wakamoto T, Kitahara R, Kanbayashi S, Takahashi H, Murata A, Kameda T. Rational design using sequence information only produces a peptide that binds to the intrinsically disordered region of p53. Scientific Reports. 2019:9:8584. DOI: 10.1038/s41598-019-44688-0

[34] Khazanov N, Levy Y. Sliding of p53 along DNA can be modulated by its oligomeric state and by cross-talks between its constituent domains. Journal of Molecular Biology. 2011:408:335-355. DOI: 10.1016/j.jmb.2011.01.059

[35] Terakawa T, Kenzaki H, Takada S. p53 searches on DNA by rotation-uncoupled sliding at C-terminal tails and restricted hopping of core domains. Journal of the American Chemical Society. 2012:134:14555-14562. DOI: 10.1021/ja305369u

[36] Leith JS, Tafvizi A, Huang F, Uspal WE, Doyle PS, Fersht AR, Mirny LA, van Oijen AM. Sequence-dependent sliding kinetics of p53. Proceedings of the National Academy of Sciences of the United States of America. 2012:109:16552-16557. DOI: 10.1073/pnas.1120452109

[37] Murata A, Ito Y, Kashima R, Kanbayashi S, Nanatani K, Igarashi C, Okumura M, Inaba K, Tokino T, Takahashi S, Kamagata K. One-dimensional sliding of p53 along DNA is accelerated in the presence of Ca(2+) or Mg(2+) at millimolar concentrations. Journal of Molecular Biology. 2015:427:2663-2678. DOI: 10.1016/j.jmb.2015.06.016

[38] Murata A, Itoh Y, Mano E, Kanbayashi S, Igarashi C, Takahashi H, Takahashi S, Kamagata K. One-dimensional search dynamics of tumor suppressor p53 regulated by a disordered C-terminal domain. Biophysical Journal. 2017:112:2301-2314. DOI: 10.1016/j.bpj.2017.04.038

[39] Itoh Y, Murata A, Takahashi S, Kamagata K. Intrinsically disordered domain of tumor suppressor p53 facilitates target search by ultrafast transfer between different DNA strands. Nucleic Acids Research. 2018:46:7261-7269. DOI: 10.1093/nar/gky586

[40] Subekti DRG, Murata A, Itoh Y, Takahashi S, Kamagata K. Transient binding and jumping dynamics of p53 along DNA revealed by sub-millisecond resolved single-molecule fluorescence tracking. Scientific Reports. 2020:10:13697. DOI: 10.1038/s41598-020-70763-y

[41] Subekti DRG, Murata A, Itoh Y, Fukuchi S, Takahashi H, Kanbayashi S, Takahashi S, Kamagata K. The disordered linker in p53 participates in nonspecific binding to and one-dimensional sliding along DNA revealed by single-molecule fluorescence measurements. Biochemistry. 2017:56:4134-4144. DOI: 10.1021/acs.biochem.7b00292

[42] Itoh Y, Murata A, Sakamoto S, Nanatani K, Wada T, Takahashi S, Kamagata K. Activation of p53 facilitates the target search in DNA by enhancing the target recognition probability. Journal of Molecular Biology. 2016:428:2916-2930. DOI: 10.1016/j.jmb.2016.06.001

[43] Takada S, Kanada R, Tan C, Terakawa T, Li W, Kenzaki H. Modeling Structural Dynamics of Biomolecular Complexes by Coarse-Grained Molecular Simulations. Accounts of Chemical Research. 2015:48:3026-3035. DOI: 10.1021/acs.accounts.5b00338

[44] Fogal V, Gostissa M, Sandy P, Zacchi P, Sternsdorf T, Jensen K, Pandolfi PP, Will H, Schneider C, Del Sal G. Regulation of p53 activity in nuclear bodies by a specific PML isoform. EMBO Journal. 2000:19:6185-6195. DOI: 10.1093/emboj/19.22.6185

[45] Guo A, Salomoni P, Luo J, Shih A, Zhong S, Gu W, Pandolfi PP. The function of PML in p53-dependent apoptosis. Nature Cell Biology. 2000:2:730-736. DOI: 10.1038/35036365

[46] Cioce M, Lamond AI. Cajal bodies: a long history of discovery. Annual Review of Cell and Developmental Biology. 2005:21:105-131. DOI: 10.1146/annurev.cellbio.20.010403.103738

[47] Kamagata K, Kanbayashi S, Honda M, Itoh Y, Takahashi H, Kameda T, Nagatsugi F, Takahashi S. Liquid-like droplet formation by tumor suppressor p53 induced by multivalent electrostatic interactions between two disordered domains. Scientific Reports. 2020:10:580. DOI: 10.1038/s41598-020-57521-w

[48] Gaglia G, Guan Y, Shah JV, Lahav G. Activation and control of p53 tetramerization in individual living cells. Proceedings of the National Academy of Sciences of the United States of America. 2013:110:15497-15501. DOI: 10.1073/pnas.1311126110

[49] Weinberg RL, Veprintsev DB, Fersht AR. Cooperative binding of tetrameric p53 to DNA. Journal of Molecular Biology. 2004:341:1145-1159. DOI: 10.1016/j.jmb.2004.06.071

[50] Hamard PJ, Lukin DJ, Manfredi JJ. p53 basic C terminus regulates p53 functions through DNA binding modulation of subset of target genes. Journal of Biological Chemistry. 2012:287:22397-22407. DOI: 10.1074/jbc.M111.331298

[51] Hamard PJ, Barthelery N, Hogstad B, Mungamuri SK, Tonnessen CA, Carvajal LA, Senturk E, Gillespie V, Aaronson SA, Merad M, Manfredi JJ. The C terminus of p53 regulates gene expression by multiple mechanisms in a target- and tissue-specific manner in vivo. Genes and Development. 2013:27:1868-1885. DOI: 10.1101/gad.224386.113

[52] Marouco D, Garabadgiu AV, Melino G, Barlev NA. Lysine-specific modifications of p53: a matter of life and death?, Oncotarget. 2013:4:1556-71. DOI:

[53] Laptenko O, Shiff I, Freed-Pastor W, Zupnick A, Mattia M, Freulich E, Shamir I, Kadouri N, Kahan T, Manfredi J, Simon I, Prives C. The p53 C terminus controls site-specific DNA binding and promotes structural changes within the central DNA binding domain. Molecular Cell. 2015:57:1034-1046. DOI: 10.1016/j.molcel.2015.02.015

[54] Retzlaff M, Rohrberg J, Kupper NJ, Lagleder S, Bepperling A, Manzenrieder F, Peschek J, Kessler H, Buchner J. The regulatory domain stabilizes the p53 tetramer by intersubunit contacts with the DNA binding domain. Journal of Molecular Biology. 2013:425:144-155. DOI: 10.1016/j.jmb.2012.10.015

[55] Friedler A, Veprintsev DB, Freund SM, von Glos KI, Fersht AR. Modulation of binding of DNA to the C-terminal domain of p53 by acetylation. Structure. 2005:13:629-636. DOI: 10.1016/j.str.2005.01.020

[56] Loffreda A, Jacchetti E, Antunes S, Rainone P, Daniele T, Morisaki T, Bianchi ME, Tacchetti C, Mazza D. Live-cell p53 single-molecule binding is modulated by C-terminal acetylation and correlates with transcriptional activity. Nat Commun. 2017:8:313. DOI: 10.1038/s41467-017-00398-7

[57] Kawaguchi T, Kato S, Otsuka K, Watanabe G, Kumabe T, Tominaga T, Yoshimoto T, Ishioka C. The relationship among p53 oligomer formation, structure and transcriptional activity using a comprehensive missense mutation library. Oncogene. 2005:24:6976-6981. DOI: 10.1038/sj.onc.1208839

[58] Rajagopalan S, Huang F, Fersht AR. Single-Molecule characterization of oligomerization kinetics and equilibria of the tumor suppressor p53. Nucleic Acids Research. 2011:39:2294-2303. DOI: 10.1093/nar/gkq800

[59] Fischer NW, Prodeus A, Malkin D, Gariepy J. p53 oligomerization status modulates cell fate decisions between growth, arrest and apoptosis. Cell Cycle. 2016:15:3210-3219. DOI: 10.1080/15384101.2016.1241917

[60] Wang YV, Wade M, Wong E, Li YC, Rodewald LW, Wahl GM. Quantitative analyses reveal the importance of regulated Hdmx degradation for P53 activation. Proceedings of the National Academy of Sciences of the United States of America. 2007:104:12365-12370. DOI: 10.1073/pnas.0701497104

www.ingramcontent.com/pod-product-compliance
Lightning Source LLC
Chambersburg PA
CBHW081228190326
41458CB00016B/5721